PHYSICAL SECURITY AND SAFETY

A Field Guide for the Practitioner

Occupational Safety and Health Guide Series

Series Editor

Thomas D. Schneid
Eastern Kentucky University
Richmond, Kentucky

Published Titles

The Comprehensive Handbook of School Safety, *E. Scott Dunlap*

Corporate Safety Compliance: OSHA, Ethics, and the Law, *Thomas D. Schneid*

Creative Safety Solutions, *Thomas D. Schneid*

Disaster Management and Preparedness, *Thomas D. Schneid and Larry R. Collins*

Discrimination Law Issues for the Safety Professional, *Thomas D. Schneid*

Labor and Employment Issues for the Safety Professional, *Thomas D. Schneid*

Loss Control Auditing: A Guide for Conducting Fire, Safety, and Security Audits, *E. Scott Dunlap*

Loss Prevention and Safety Control: Terms and Definitions, *Dennis P. Nolan*

Managing Workers' Compensation: A Guide to Injury Reduction and Effective Claim Management, *Keith R. Wertz and James J. Bryant*

Motor Carrier Safety: A Guide to Regulatory Compliance, *E. Scott Dunlap*

Occupational Health Guide to Violence in the Workplace, *Thomas D. Schneid*

Physical Hazards of the Workplace, *Larry R. Collins and Thomas D. Schneid*

Physical Security and Safety: A Field Guide for the Practitioner, *Truett A. Ricks, Bobby E. Ricks, and Jeffrey Dingle*

Safety Performance in a Lean Environment: A Guide to Building Safety into a Process, *Paul F. English*

Security Management: A Critical Thinking Approach, *Michael Land, Truett Ricks, and Bobby Ricks*

Security Management for Occupational Safety, *Michael Land*

Workplace Safety and Health: Assessing Current Practices and Promoting Change in the Profession, *Thomas D. Schneid*

PHYSICAL SECURITY AND SAFETY

A Field Guide for the Practitioner

Edited by

Truett A. Ricks • Bobby E. Ricks • Jeff Dingle

CRC Press
Taylor & Francis Group
Boca Raton London New York

CRC Press is an imprint of the
Taylor & Francis Group, an **informa** business

CRC Press
Taylor & Francis Group
6000 Broken Sound Parkway NW, Suite 300
Boca Raton, FL 33487-2742

First issued in hardback 2017

Version Date: 20140523

ISBN-13: 978-1-4822-2702-4 (pbk)
ISBN-13: 978-1-138-43090-7 (hbk)

Library of Congress Cataloging-in-Publication Data

Physical security and safety : a field guide for the practitioner / edited by Truett A. Ricks, Bobby E. Ricks, Jeffrey Dingle.
pages cm
Summary: "This book addresses the need for security, how to identify the threat against and level of security needed to protect assets, and security theories and concepts used to provide an acceptable level of security. Security practices are intermingled with safety issues. OSHA and Fire Regulations complement an overall loss prevention program. This text introduces the Security practitioner to basic tenets of safety and safety management. Safety practitioners will find this book of value in understanding physical security theories, concerns, and practices"-- Provided by publisher.
Includes bibliographical references and index.
ISBN 978-1-4822-2702-4 (paperback)
1. Buildings--Security measures. I. Ricks, Truett A. II. Ricks, Bobby E. III. Dingle, Jeffrey.

TH9705.P53 2014
658.2'8--dc23 2014020108

Visit the Taylor & Francis Web site at
http://www.taylorandfrancis.com

and the CRC Press Web site at
http://www.crcpress.com

Truett A. Ricks
1935–2014

As we complete this book, we mourn the passing of Dr. Truett Ricks. This is his seventh book, with six prior books related to physical security and safety. Days before his death, he was collaborating with authors to finalize their chapters and to complete this book. An inspiration to all who knew him, we dedicate this book to his legacy.

Bobby Ricks
Coauthor and son

Contents

Preface

This is a how-to guide written by practicing professionals. The idea for this book is to provide basic yet essential knowledge on facility security measures and how safety considerations interact with security. For the security manager, the book will act as a handbook for security applications with key safety considerations. For the safety manager, the book will act as a handbook of key safety considerations and a reference of security considerations. For the facility manager, the book provides fundamental information for a better understanding of security needs.

The book is broken into two parts: "Security and Safety Planning" and "Infrastructure Protection."

"Security and Safety Planning" begins with the theory and concepts of security to explore the why of security needs. Threat decomposition, risk assessment, and surveys begin to identify security threats and vulnerabilities so the manager knows what to protect, how to protect it, and how much risk the organization is willing to take to *not* protect organizational assets. The survey process will also identify general safety concerns and specific issues covered by Occupational Safety and Health Administration (OSHA) and fire protection regulations. A survey instrument is included in the book's appendix. The part ends with a discussion of security policies and procedures for implementing a system and developing an attitude of effective physical security.

Part II begins by looking at the overall physical protection program. Access control, perimeter control, and alarm systems are examined, with information on security lighting and closed-circuit television to complement the physical protection program. Response force models are examined for when circumstances demand an on-site response team. This part concludes with practical considerations for protecting information technology (IT). IT security is explained in terms common to the nontechnical manager and discusses nontechnical measures the security manager can implement to protect IT systems.

About the Authors

The professionals writing this book have years of experience in security and loss prevention, safety, fire protection, law enforcement, homeland security, and law.

Jarred Ball, MS, is a certified law enforcement instructor in homeland security for the Commonwealth of Kentucky. He has a master's degree in safety, security, and emergency management from Eastern Kentucky University and has taught homeland security at the college level. Ball has conducted antiterrorism risk and vulnerability assessments of critical infrastructure for the Commonwealth of Kentucky's Office of Homeland Security.

Jeff Dingle, CPP, is the director of security for PCI Gaming in Atmore, Alabama. He is a former special agent with the National Security Agency, and director of security for the Carter Presidential Library and Home Depot. He has written and taught physical security courses for over 25 years for private companies and the Federal Law Enforcement Training Center (FLETC), being the program manager for the Advanced Physical Security Training Program and the Physical Security Managers Training Program. He has written numerous articles on security and has presented at industry security conferences.

James L. Pharr, MS, is an associate professor of fire and safety engineering technology at Eastern Kentucky University (EKU). Professor

Pharr specializes in fire dynamics, building and life safety, leadership, emergency scene operations, and fire investigation. Prior to joining EKU, Pharr was the emergency management director and fire marshal in Gaston County, North Carolina.

Bobby E. Ricks, JD, is an associate professor at Midway College. He has written books on security and management, and has been teaching security for over 25 years for private companies and for the FLETC. He is a former special agent with the Federal Bureau of Investigation, former director of Crime Prevention for the Richmond, Kentucky, Police Department, and was a member of the U.S. Air Force Security Forces. He has consulted with numerous law firms on security matters.

Truett A. Ricks, PhD, was the retired dean of the College of Justice and Safety at Eastern Kentucky University, and wrote several textbooks on physical security and crime prevention. He was the former owner of a security company and was president of his own consulting firm. He was a certified protection professional (CPP).

Truett Graham Ricks, MBA, works as an IT systems analyst specializing in IT security and the creation and review of IT-related policies. He has experience with a wide range of systems and clients.

Thomas D. Schneid, JD, LLM, is the chair of the Department of Safety and Security in the College of Justice and Safety at Eastern Kentucky University. Tom has worked in the safety and human resources fields for over 30 years at various levels, including corporate safety and industrial relations. In Tom's legal practice, he has represented numerous corporations in OSHA and labor-related litigations throughout the United States. He has authored or coauthored numerous texts on safety and labor law.

Thomas Woodall Sr., is the president of Security Answers, LLC. He is the retired director of the Federal Protective Service for the U.S. General Services Agency. He has conducted numerous security surveys in government and private facilities. He has taught physical security for over 25 years for private companies and at the FLETC, being part of the Curriculum Development Committee for the Advanced Physical Security Training Program at FLETC.

PART I

Security and Safety Planning

1

THEORY OF SECURITY

JEFF DINGLE

Contents

Many textbooks date the origin of security back to ancient Egypt or to Europe after the fall of the Roman Empire. The modern era of security began September 11, 2001. Through the years, the United States has remained mostly immune from acts of terrorism, but the single horrific events of 9/11 have shown us the tragic results of what can happen. The events of September 11 caused us to totally rethink our approach to security operations. While many of the threats have not changed, new threats have been added. Today's threats include criminal acts from a variety of sources, both internal and external to an organization. Security operations may protect from a variety of things—theft, arson, acts of violence, vandalism, espionage, domestic violence, and ever-irate customers. Everyone protects what they hold valuable. Homeowners protect their homes, businesses protect their businesses, schools protect against theft and violence, the government protects everything. What varies is the amount of effort that goes into protecting what you have.

Although there have been many different theories, and different people involved, since the beginning security has basically been the same. We look for problems or potential problems, and we look for solutions to those problems. The difficulty is that we look not only for existing problems, but for potential problems that have not yet occurred, so we can take steps to prevent them from happening. It is very difficult in a business environment to spend money to prevent things that have never happened. But that's what we do. It is often better to be proactive and spend a little money now than to be reactive and wait for a problem to occur.

Security is the creation of a circumstance that is problem-free, or a circumstance that results in a minimum of problems. How a person views the need for security is affected by his or her perception of the current level of security and his or her perception of the need for security. Different people, with different backgrounds, different training, and different experiences, perceive the need for security differently. Part of the perception of need is the perception of how valuable what you have is, and how valuable what you have is to a criminal. From a purely criminological perspective, a criminal will base his or her decision on whether or not to commit a crime on several factors. One factor is the "reward" from the crime—or what he or she gets out of committing the crime. The other factor is the risk of getting caught. For years, security professionals based the level of security on that combination of risk versus reward. The greater the potential loss, the more that was spent on security. We believed that we could deter a criminal from committing any crime by simply increasing the risk. Risk was increased by increasing the level of security and making it more difficult to commit the crime, a simple process. The amount of security provided was directly correlated to the value of what was being protected. Locks could be added or improved. Fences could be built. Cameras could be installed. Risk versus reward changed with the introduction of terrorism into the United States. A terrorist simply doesn't care about the risk. The motivation to commit a crime is different, and an increase in security is not a deterrent. The old model of risk/reward no longer is valid, and we have to consider not only what kind of potential value we have to a criminal, but also what potential value we have to a terrorist.

Security specialists now have to consider how they handle any incident differently. A great example is the Murrah Building in Oklahoma City. As a federal building, multiple security surveys over the years never indicated a serious threat. A truck bomb and two terrorists destroyed the building, killing 168 people. This shows, simply, that terrorism can occur anywhere, at any time. But terrorism isn't our only threat. Acts of violence can occur anywhere. Heritage High School in suburban Atlanta was the site of a school shooting in May 1999, less than a month after the school shooting in Columbine, Colorado. There have been a growing number of incidents of school violence across America, and this shows again that the need for security exists everywhere. We need to change our thinking about security.

What Are You Trying to Accomplish?

A key to any successful security operation is identifying your ultimate goal. This is a question that will be asked over and over again in this text. Knowing what to expect, and what not to expect from cameras, access control, intrusion detection, and other security options is critical to creating and maintaining a secure environment. Security successes are not always easy to see. In a speech shortly after the September 11 bombing, then President George W. Bush said, "Some victories will be won outside of public view, in tragedies avoided and threats eliminated" (National radio address, September 29, 2001). This quote helps to explain what we are trying to do, what we are trying to accomplish in any security operation. It is sometimes difficult to quantify the success of a security organization. In an effort to accomplish better security, the Department of Homeland Security combined 22 different federal departments and agencies into a unified, integrated cabinet agency when it was established in 2002.*

In reality, we protect everything. You write your name on your lunch bag. You lock your car when you park it. The problem is that "government cannot protect all possible targets for every conceivable kind of attack" (Brian Jenkins, the Rand Corporation).† As a result, we have to make choices. With limited budgets and resources, we have to choose what we will protect and what we will not protect. These can be difficult choices. How do we choose?

Some elements of this decision making are obvious. We provide the greatest protection to the things that are most valuable to us. While that seems easy, it is often difficult to determine what is, in fact, most valuable. As before, value can be based on individual perception. However, some of our techniques are based on common sense, and we often base our planning on the existing threat level.

We use a fairly simple process to determine the ideal security level. We look at where we are, assessing the current levels of security equipment and policies. Then we look at where we need to be; where we need to be includes the security and equipment, policies,

* www.dhs.gov/history

† David Shribman, The Textbook Approach to Terrorism, *New York Times*, April 22, 1984, retrieved from http://www.nytimes.com/1984/04/22/weekinreview/the-textbook-approach-to-terroism.html

and procedures that are not currently in place, but are necessary. The third step is to determine how to get from where you are to where you need to be. These three steps are accomplished by utilizing a risk assessment and conducting a security survey. Risk assessments and security surveys are addressed later in this course.

When looking at an overall security program, there are two very different approaches, proactive and reactive. Being proactive involves looking for problems before they occur, and taking steps to prevent problems before they happen. Being reactive involves waiting for a problem to occur, and only addressing issues after they occur. Generally speaking, law enforcement is reactive. The police wait for a crime to happen, and then take steps to solve the crime. Generally speaking, security operations are proactive. Security operations involve looking for potential problems and taking steps to eliminate the problems before they occur. From a business standpoint, reactive is much easier to justify. Solving problems that have occurred is easy to quantify. You can judge the success of a police officer by things you can count—number of tickets written, number of arrests, number of convictions. It is difficult to quantify a successful security program. While you can look at trends in general, it is hard to determine how many crimes did not occur due to the security measures put into place.

The Crime Triangle

From a criminological standpoint, three things must occur in order for a crime to be committed. Similar to the fire triangle, all three things must occur at the same time in order for a crime to occur. First, a criminal must be motivated to commit a crime. Many things may motivate an individual to commit criminal acts. Motivation may be hard to define, and although we try with incarceration, it is difficult to remove motivation from someone who is intent on committing a crime. Second, a criminal must have the tools and means to commit a crime. In today's society, tools are easy to obtain. Tools that cannot be stolen can be easily purchased on the Internet, or they can be fabricated. Techniques that were once difficult to learn are now easily found on the Internet. Websites like TOTSE.com and lockpicking101.com bring knowledge and skill sets to people that would otherwise never gain them.

Twenty years ago, if you were 15 years old and wanted to learn how to make a bomb, options were available, but limited. You could buy, borrow (or check the library) the *Anarchist Cookbook*, or order it from the back of *Soldier of Fortune* magazine. The information was available, but you had to work for it. Today, an Internet search for "make a bomb" returns 614,000,000 hits. A criminal's two most dangerous tools are his or her hands and mind, something we cannot take away from him or her. Last, the third requirement of the crime triangle is opportunity. Much of crime is based on opportunity. Of motivation, tools and means, and opportunity, opportunity is the only point where we can really have any impact. Much of a good security program involves removing opportunity. So what happens when we remove opportunity? Generally, one of four things occurs.

Sometimes, crime stops. Sometimes, criminals simply find it too difficult to commit a crime, and stop. This is uncommon. More often, criminals will simply wait until a better opportunity presents itself. Sometimes, criminals commit different crimes. If there is too much security in place to rob a jewelry store, it might be simpler to steal a car. Recently, a midwestern state announced that it was showing a tremendous impact on stopping meth labs—so much success that there was a marked increase in theft. People unable to produce meth had to go back to the old ways of stealing. Most often, criminals simply move to other locations where there is more opportunity and less security. This is known as displacement. Displacement is the most common result of an increased security program. Criminals will follow the path of least resistance and look for easy targets. It is important to understand displacement because you should be aware that while you are trying to push crime away from your facility, others are trying to push crime away from their facility. Your facility does not have to be the best secured in the world, in your state, or even in your city. However, your security must be equal to or better than the security of facilities immediately surrounding you.

Security impacts our daily life, often in ways we no longer recognize. Security has a direct impact on convenience. As a rule, as security increases, convenience is impacted. Security procedures are often inconvenient, however necessary. Airports are the best example of convenience—when traveling, we take off our shoes for security reasons. Access control is often inconvenient and slows the access process. Likewise, security can have a direct impact on aesthetics. In any facility, a decision must be made as to the importance of visible security devices. It is possible to create a secure environment and maintain aesthetics, but generally speaking, making security pretty is expensive. Security design is addressed later in the text.

There is much to address in how security operates today. A primary career in security presents challenging opportunities, while many business operations demand knowledge of and proficiency in security operations.

2

Concepts of Security

BOBBY E. RICKS AND JEFF DINGLE

Contents

Security Design

Security begins with security and safety in mind. Security design has the goal of making the public and employees feel safe and secure while increasing the perception of risk to an attacker. Design considerations must take into account aesthetics while providing a level of security. David Childs, architect for the new One World Trade Center, remarked: "One of the things I have learned about safety is the appearance of safety. If somebody looks at a building and says, 'Boy, that looks pretty flimsy,' they are interested in it. If it looks tremendously strong, you say, 'Well, let's go pick on somebody else.'" His thoughts are on looking and being safe. You can provide a sense of openness to a facility while providing a level of security to deter attackers.

In the business world, security is a cost that does not produce revenue. Budget restrictions may dictate the amount of security a facility may provide. The security manager must take into consideration the risk and work to provide the best level of security for the greatest risks. Operational efficiency plays into the equation. Convenience is sacrificed as you increase security. Where one once walked unhindered, security measures dictate greater access control, which in turn slows down throughput. A jewelry store or bank that requires customers to wait before they are allowed in the facility may find that their customers choose to do business elsewhere.

Design must consider operational needs while providing a level of security at an acceptable cost to deter a broad spectrum of threats. Traditional means were to install intimidating and unsightly fencing, access control and alarm systems, and barriers and lighting. Crime prevention through environmental design (CPTED) uses nontraditional means to provide higher levels of security that is often invisible. The beautiful landscaping between the street and the building may provide security against heavy vehicles while giving the impression of greenery and aesthetics. The earth surrounding the trees may be firm enough for a person to walk on, but collapse under the weight of a heavy vehicle. CPTED is used by architects and designers to control behavior. The hard seats, bright colors, and bright lights at a fast food restaurant differ from an upscale restaurant with padded seats, earth tones and pastels, and dim lights. One is easy to clean and encourages a "get in and get out" traffic flow, while the other encourages a relaxed atmosphere.

CPTED actors involve normal users, those who work and shop in a facility; abnormal users, those who may intrude in or on land for their convenience; and observers. For example, a convenience store is designed to get the patrons to buy gas and other items while keeping people from using the site as a shortcut, and offering an open view for passersby to create a sense of observation that is a deterrent to attackers. A bus stop shelter bench can serve as a comfort to patrons, while sturdy design can serve as a barricade that helps protect a facility. The bench can be designed for short-term comfort and to prevent people from staying too long or sleeping on it. Decorative planters can replace unsightly bollards yet still provide a high measure of security.

Attacks on facilities force the implementation of measures to prevent the loss of government resources and allow the continuation of services in the event of a mishap. This requires planning and preparation of security measures to assure the protection of government personnel and property. The amount of risk an organization is willing to bear depends on the probability of an event occurring and the severity of impact of that event on the organization's operations. Security risk management involves identifying the organization's assets and the vulnerabilities to those assets. The level of protection necessary to safeguard those assets is measured against the actual level of protection provided. The discrepancy between the desired level of protection and actual level of protection is the risk. There is more detailed information on risk assessment in Chapters 4 and 5.

Emergency Preparedness

Disasters happen, both natural and man-made, and security managers need to plan and prepare for common disasters. The response during the first few hours of a disaster has a great impact on the morale and recovery of the organization. A fast, effective response can save lives, prevent injuries, and prevent or minimize property damage. Emergency management involves participants in the private sector and at all levels of government: local, state, and national. Disaster response activities are geared according to phases before, during, and after emergency events. The effectiveness of emergency management rests on a network of relationships among partners in the system. Security professionals should partner with and coordinate their emergency planning with local emergency management agencies as well as neighboring businesses and industry to know what they can expect from local and state emergency services in the event of a disaster. For a private business, downtime is lost revenue. Effective planning for common disasters gets the organization operational in an efficient and effective manner. Security personnel should be trained by emergency management personnel in what to do before and during a disaster, as well as their role in recovery after a disaster has occurred. The Federal Emergency Management Agency (FEMA) has free online courses personnel can take to familiarize themselves with emergency response and recovery procedures and protocols.

Loss Prevention

The majority of theft and inventory shrinkage comes from within the organization. Loss prevention measures use policies and procedures to minimize the risk of loss. The adage "locks keep honest people honest" describes removing the temptation to steal or intrude in a private area. Loss prevention measures promote accountability as well as keeping a watchful eye on your assets. Inventory and invoice controls reduce the opportunity for stock to go missing. Just-in-time inventory techniques reduce the amount of stock on hand, which helps monitor stock. A high-volume auto dealer wasn't aware a former employee had stolen a car until the police went to arrest him on an unrelated matter and found an unregistered vehicle at his house. Loss prevention uses common sense and proven techniques to deter unwanted activity. For example, a convenience store with two employees on duty is extremely less likely to be robbed than a store with one employee only. Business opening and closing should also be with two people. Rotating employees reduces collaboration between employees on dishonest practices. Skimming registers periodically reduces the amount of cash taken in the event of a robbery. Low rewards discourage robbers and deter further activity. Time-delay safes allow employees to drop money into a safe that can be retrieved later if change is needed. A timer delays the retrieval, which discourages would-be thieves. For example, if a cashier needs to replenish $5 bills, he or she requests a "drop" that may take 10 minutes to complete. Robbers aren't going to wait for the drop. When taking money to or from the bank, vary your times and disguise deposits. Bank runs are best with two people, or at a minimum, have another employee escort you to your car and have a bank employee meet you outside the bank.

Place items at risk for shoplifting where they are visible by cashiers and other store personnel. Training clerks on shoplifting techniques can reduce losses simply by having observant clerks looking out for suspicious activity.

Legal Concerns

Lawsuits for security deficiencies have increased in the past 20 years, with judgments holding property owners liable for injuries and loss due to improper or inadequate security. The law recognizes no general duty to protect or come to the aid of another. The imposition of duty, defined as reasonable affirmative steps to aid a person in peril, arises in many relations to include owners of premises held open to business visitors. Courts will generally find a duty where reasonable persons would recognize it and agree that it exists. Duty may arise from statute, regulation, or case law.

Statutory duty exists when federal, state, or local laws set forth legislation requiring or authorizing the protection of assets. For example, Title 40 USC § 318(a) authorizes the General Services Administration to establish rules and regulations regarding the protection of federal property. An example of a regulation is 41 CFR § 102-20.301, authorizing inspections of packages on federal property.

Court decisions, also known as case law, apply common law concepts where there is no statutory law to create a rule of law to be followed in a particular situation. Case law may create a duty to protect even when no statute or regulation establishes such a duty. Case law on duty to protect varies from state to state. The duty to protect extends to persons on the property for an official purpose, known as licensees or invitees. A licensee is one on the premises for his or her benefit, such as a customer or client. An invitee is one on the premises for business purposes, primarily for his or her benefit, such as consultants or vendors. In one case, courts held that one who comes on the premises to use the public restroom is an invitee. This applies to any facility open to the public, whether it is a private or a public facility. Invitees and licensees are owed a standard of care to protect them from hazards on the property that are known or reasonably could be identified. For example, retail stores know that customers entering the store with wet umbrellas will drip water on the floor, causing a hazard. They take reasonable steps, such as giving customers plastic bags to store their umbrellas, patrolling the showroom floor looking for wet spots to dry up, and placing warning signs to alert customers.

A third category of person is the trespasser. This person is not authorized to be on the property. While the trespasser is not owed the same duty of care as invitees and licensees, he or she is owed a duty of protection against inherently dangerous activities. Some inherently dangerous activities are called an attractive nuisance. These are things that may cause a person to trespass. Swimming pools at a private residence would be an attractive nuisance. Because of the fear of children drowning, local ordinances and even homeowners insurance may require a protective fence around the pool. While a residential example is used, the same legal concept applies to commercial and government properties as well.

Protecting persons on the premises from attacks from strangers or employees is not the same as securing an area against outside attacks.

Security measures to guard against outside threats will normally afford some protection to those on the property. On such a facility, one that is secured against outside attacks, the threat of attacks against invitees is less but ever present.

When courts find a duty to protect, the owner/possessor of a facility must take reasonable steps to reduce the risk of reasonably foreseeable crimes committed on the premises. Some jurisdictions extend foreseeability to areas off the premises. Extending a duty to protect off the premises usually arises when the area off the property is somehow connected to the use of the property, such as a parking lot. Foreseeability of crimes is generally determined by case law. It is a defense to the duty to protect where the act was not foreseeable.

A reasonable standard for determining foreseeability of crimes is to look in the "building neighborhood" for crimes that have the potential of occurring on government property. The U.S. General Services Administration considers this area to be 10 city blocks.

Natural or artificial boundaries, like a river or an interstate highway, may alter the 10-city-block rule. The boundary makes it difficult for activity on one side of the boundary to influence activity on the other side of the boundary. This will have an impact on the severity of incidents in your area.

The severity and frequency of crimes will direct the type and level of security necessary to provide adequate protection. Local police should be contacted to obtain statistics on criminal activity in your area. If available, obtain information from dispatch records, as some police calls may not warrant an official police report, but dispatch records would provide a better picture of the building neighborhood.

Law enforcement agencies normally don't hide information, especially from other agencies or security personnel. Politics may prevent certain information from being made public so as not to scare away a potential business or industry from locating in the area. Reports through the grapevine from other security practitioners and professionals in your area will help in assessing what type of protection is necessary.

Adequacy of security is measured by looking at the level of security provided by other activities in the area that are similar to yours.

This is determining the “custom” of an area. Custom is a community standard, meaning the usual and customary conduct of others under similar circumstances. Looking at other facilities in the area that are similar to the facility in question, and meeting the standard they have set, would be what the security community regards as proper.

Community standards will change according to the facility being protected. Office buildings, banks, government offices, shopping centers, and other industries in a community will be measured against like industries for the community custom. A nuclear power plant would not necessarily be measured against a hydroelectric power plant.

A community may change according to the standard. Every nuclear power plant in the country may be “the community.” This community would be the measuring stick of the custom for that industry. This is because of the small number of nuclear power plants in the United States and the highly regulated nuclear power industry. Hydroelectric power plants could be measured against other hydroelectric plants in a region of the country since there are more hydroelectric plants and they are not as heavily regulated as nuclear power plants.

To establish a baseline of security, surveys should be conducted at regular intervals, or whenever there is a significant change in the facility, neighborhood, or crime. A significant change would be anything that would have an impact on the validity of the prior survey. An example would be where a new tenant with sensitive resources moves into a building or a low-income housing project opening across from your facility.

The surveyor should have sufficient credentials to establish his or her qualifications to conduct surveys. Qualifications for security surveyors will include the skills and knowledge of the surveyor as measured by his or her experience, education, and training. Any law enforcement or security-related experience should be noted, as well as basic and advanced law enforcement and security training programs, college courses, or degrees. This will be the foundation for establishing the surveyors as “experts” in the field of security.

When a lawsuit claims inadequate security, the surveyor will be called to testify and attest to the need for a certain level of protection. Convincing the trier of fact, whether that be a judge or a jury, is up to this person.

Before reducing security, a survey should be completed showing justification for the reduction. Any reduction should be justified by showing a reduction in crime in the area or a change in the facility that would warrant the change.

Liability for Employee Conduct

Courts require due diligence in the hiring process to protect employees and visitors from an unfit employee. An employer may be liable for the acts of an employee when it knew, reasonably should have known, or could reasonably ascertain an employee was unfit for the position or knows that hiring or retention of the employee could create a risk of harm to others.

3

Threat Decomposition

BOBBY E. RICKS AND THOMAS D. SCHNEID

Contents

As a security manager, assessing the potential threats is essential to providing adequate security for your facility. Threat decomposition is about identifying threats most likely to affect your operation. Potential actions include theft, pilferage, espionage, sabotage, bombing, arson, compromise of data or trade secrets, kidnapping or assassination of key personnel, disruption of operations, and embarrassment. Liaison with law enforcement and other security managers in your area will help identify some risks and the likelihood of problems occurring from those risks. Most operations have limited resources, and the prioritization of threats allows you to address concerns most detrimental to the operation and to justify the cost of protection. Even with unlimited resources, threat decomposition helps you plan the order in which your threats will be addressed.

Start by identifying specific assets that are potential targets in your organization and the threats to those targets. For example, a government contractor with a security clearance may identify classified information as a target of interest to others. Potential threats would be hostile governments, terrorists, and organizations opposing the technology or resource. Once a potential threat is identified, categorize the threat by identifying if the threat is skilled, unskilled, or semiskilled. Skilled threats target sensitive and high-value items, and conduct military-style operations such as espionage, sabotage, or theft. They have extensive technical and financial support for their operations, and will work to gain extensive knowledge of the site, your security posture, and can identify vulnerabilities. They conduct thorough planning, and may have someone working inside or have gathered information on private areas of a facility through social intelligence (getting to know someone inside and surreptitiously asking questions that reveal information, such as locks, alarms, closed-circuit televisions [CCTVs], and other security details). Skilled threats can compromise security systems surreptitiously and can enter and leave no evidence of their attack.

Unskilled threats are usually opportunistic with little or no planning or sophisticated equipment. Their operations are for monetary gain or revenge on a current or former employer. Knowledge of security systems is very limited, and the ability to compromise security systems is crude, usually resorting to vandalism to defeat security measures. In between these two are semiskilled threats that are aware of some security measures and plan to make use of limited resources. Technical support comes from open sources. The semiskilled threat is usually for personal gain or embarrassment.

Once the level of threat is identified, brainstorm possible scenarios of groups, their skill level, and their level of motivation to obtain or compromise the targeted asset. Tabletop exercises are useful in playing out a scenario to determine the effectiveness of existing security measures to protect the asset. From the information gleaned from these exercises, one can begin to justify the current security posture and make recommendations to improve security.

Terrorism

Terrorism has many definitions, usually written to address a specific concern to political, military, or law enforcement interests. A general definition of terrorism is: violence or the threat of violence calculated to create an atmosphere of fear and alarm to bring about social change. From this base, you can identify potential threats against your organization as terroristic or nonterroristic. Terrorism differs from classic crimes, as it is executed with the deliberate intention of causing panic, disorder, and terror in an organized society. Terrorists, by their acts, attempt to inspire and manipulate fear for a variety of purposes—embarrassment, disruption of services, financial gain, etc.—with the ultimate purpose of furthering their cause. Attacks are choreographed to attract the attention of the media, which is a force multiplier. The attacks are aimed at the people watching, not the actual victims.

In order for a terrorist group to justify their violence, and for individual terrorists to participate in violent acts, they must be committed to a belief and feel they are serving the greater good in committing these acts. This is social justification. A person feels approval from the group in advocating violent action for change. People who are attracted to terrorist groups are generally social outcasts. They have been rejected by society and tend to group with like-minded people. Sometimes the need for acceptance causes someone to identify with a group just for social acceptance. Antisocial behavior is rewarded by the group, and the person is motivated to continue such conduct. Psychological justification comes from the group identifying an enemy and creating an atmosphere of "us against them." The group must have an issue that drives them to violent action for social justice. They "demonize" the enemy and look for reasons to distrust the enemy, reinforcing their cause and conduct in the process.

While ideological terrorism is waning, groups motivated by religion are growing. Religious terrorists act differently than ideological or single-issue terrorists. They are not constrained by the political or cultural norms. They don't seek to kill their enemies; they are out to remove evil. Terrorists can be classified into three groups: criminals, who terrorize society for monetary gain or revenge; crazies, who are motivated by the thrill of violence or the feeling for power; and crusaders, who feel they must be violent for society to change for the better. Knowing which group your threat matches helps plan for possible activity. Motivation—religious, ethnic, social, or antigovernment—helps assess their level of commitment to succeed in their mission to steal, destroy, or compromise the asset you are protecting. While your asset may not be the target for a potential group, resources on your property may help them in their plight to achieve their ultimate objective.

When a potential terrorist group is identified in your area, or with an interest in your assets, determine the likelihood of an event by the history of the group's activities in your area. Identify the capability and motivation of the group to commit their acts, and target intelligence that may indicate a potential act against your organization. Coordination with law enforcement and other security practitioners will broaden your network of intelligence to protect your assets.

Gangs

A gang is three or more persons who have a common identifying sign, symbol, or name that individually or collectively engage in criminal activity. A gang acts like a terrorist organization, but doesn't always fit the terrorist definition, and the term *gang* is more specific to activity and conduct distinct from terrorist activity. According to the National Gang Intelligence Center (NGIC), 48% of violent crime and up to 90% of violent crime in some cities are from gang activity. Assault, threats, and vandalism top the list of high gang involvement in crime, with theft, robbery, and burglary topping moderate involvement in crime. Moving from traditional gang-related crime such as weapons and drug trafficking, gangs are moving into white-collar crime. From a physical security perspective, your threat decomposition on gangs will usually focus on local gang activity that may see your facility as a target of opportunity.

Street gangs are divided into national street gangs and neighborhood-based gangs, the latter not having an affiliation with a national group. Neighborhood-based gangs are ranked first in significant threats to criminal conduct.

The average gang member is 12–24 years old and can come from low- or upper-level socioeconomic families; some attend college, and others have military experience. People join gangs for protection, respect, and power, to make money, and for parties and drugs. Some children grow up in a neighborhood where joining a gang is almost a way of life. Problems at home may make children prefer the gang as a family who cares and looks out for their well-being.

Gangs usually claim a particular area of a town, called turf, and aggressively defend it against other gangs. In addition to territory, gangs are organized around race, ethnicity, and moneymaking activities. Gang members wear particular items as a type of uniform to distinguish themselves from others and to recognize one as being part of a gang. They may have signs or handshakes to identify their membership, and turf is often marked by graffiti. Gang recognition also includes tattoos, hand signs, codes, nicknames, and slang terms. Local law enforcement can assist in identifying gang identifiers and graffiti in or around your property and providing information on the gang's activities.

While gang graffiti is only about 15% of all graffiti, people consider all graffiti as bad and act accordingly; they shop in areas with no graffiti and fear coming to work in an area infested with graffiti. Graffiti is the largest act of vandalism in the United States. Graffiti impacts retail sales where it is found and heightens the fear of gang activity and a downward spiral of the community. The broken windows theory, a term coined by James Q. Wilson and George L. Kelling, is a concept that a broken window, left unattended, gives the appearance that no one cares, leading to acts of vandalism, which leads to more violent crimes. William Bratton was hired to head the New York transit police and embraced the theory. He took the broken windows theory and applied it to practice, cleaning up graffiti as it occurred. Graffiti artists looked for other places to ply their art.

Estimated costs of graffiti cleanup are $1–3 per person annually for community cleanup. This does not include cleanup by private business and industry. Rapid removal and cleanup of graffiti exponentially lessen reoccurrences (see Figure 3.1). Private industry should work with community leaders to enforce antigraffiti and vandalism laws to deter illegal conduct.

Gang recognition is also found in tattoos. Tattoos are popular, and it's not uncommon for potential employees to have one or more tattoos, which may be innocent, although they are recognized as a connection to a gang. NGIC reporting indicates gang members in 57 jurisdictions have applied for or gained employment with judicial, police, or correctional agencies. Human resources personnel are familiar with hiring and discrimination laws and should work with security to identify tattoos and conduct a thorough screening before hiring.

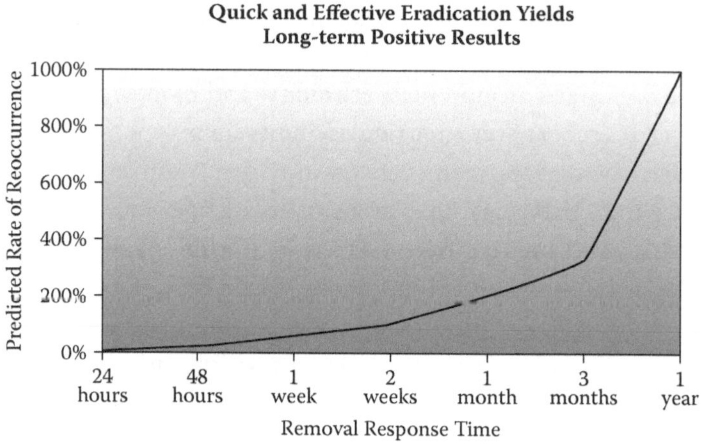

Figure 3.1 Quick and effective eradication yields long-term positive results. (Source: Graffitihurts.org).

Outlaw Motorcycle Gangs

Outlaw motorcycle gangs (OMGs) are criminal organizations whose members engage in criminal activities such as violent crime, weapons trafficking, and drug trafficking. OMGs rank first nationally as a moderate threat to criminal conduct, slightly ahead of neighborhood-based gangs and national-level street gangs. There are more than 300 active OMGs within the United States, ranging in size from single chapters with five or six members to hundreds of chapters with thousands of members worldwide. OMGs pose a serious threat to social order and are strongly linked to drug trafficking and cross-border drug smuggling. Transnational OMGs coordinate drug smuggling operations in partnership with major international drug trafficking organizations (DTOs).

Organized Crime

Organized crime (OC) is a continuing criminal enterprise that profits from illegal activities, usually activities in demand, such as gambling and prostitution. Traditional OC families used terrorist-style tactics to extort protection money from businesses, or to force a business or industry to use their services such as trash collection, concrete or paving, and even construction. Theft and truck hijacking were common OC operations. Later, groups moved into drugs and other high-return ventures, often running scams and money laundering through legitimate business interests. Federal Racketeer Influenced and Corrupt Organizations (RICO) statutes were enacted to address these illicit interests. Recent prosecution and conviction of key figures in Italian/Sicilian La Cosa Nostra (Mafia) families have curbed major operations.

While La Cosa Nostra does not pose as serious a threat as in the past, security managers need to be aware of groups in their areas that may pose a threat to theft, hijacking, or attempts to bribe or control employees or services. Other OC groups from Russia, Mexico, Asia, and Europe pose a threat to businesses operating abroad, especially in areas of executive protection. Some groups operate in the United States and are involved in theft, money laundering, and other crimes that may target your facility and operation. Security managers need to be aware of the groups operating in their areas and identify the potential threat from such groups.

Workplace Violence

Security professionals should be aware that workplace violence continues to be a major area of concern for employers as well as employees in the American workplace. Beyond the proactive measures that security professionals take to prevent workplace violence in their operations, security professionals should be aware that although the Occupational Safety and Health Administration (OSHA) does not have a specific standard addressing workplace violence, it can and will cite employers for hazards related to workplace violence under the General Duty Clause of the OSH Act. Security professionals can find information regarding workplace violence on OSHA's website, located at www.osha.gov, as well as the National Institute for Occupational Safety and Health (NIOSH), located at www.cdc/niosh.gov, and other sources.

What Is Workplace Violence?

Workplace violence is any act or threat of physical violence, harassment, intimidation, or other threatening disruptive behavior that occurs at the work site. It ranges from threats and verbal abuse to physical assaults and even homicide. It can affect and involve employees, clients, customers, and visitors. Homicide is currently the fourth-leading cause of fatal occupational injuries in the United States. According to the Bureau of Labor Statistics Census of Fatal Occupational Injuries (CFOI), of the 4,547 fatal workplace injuries that occurred in the United States in 2010, 506 were workplace homicides. Homicide is the leading cause of death for women in the workplace. However it manifests itself, workplace violence is a major concern for employers and employees nationwide.

Workplace violence falls into one of four types:

1. Criminal intent: Acts by people who enter the workplace to commit a crime.
2. Customers/clients/patients: Violence directed at employees by customers, clients, patients, students, or others to whom the employer provides a service.
3. Coworker violence: Violence against coworkers, supervisors, or managers by current or former employees.
4. Personal: Violence by someone not in the workplace, but known to or having a personal relationship with an employee.

Who Is at Risk of Workplace Violence?

Nearly 2 million American workers report having been victims of workplace violence each year. Unfortunately, many more cases go unreported. The truth is, workplace violence can strike anywhere, anytime, and no one is immune. Research has identified factors that may increase the risk of violence for some workers at certain work sites. Such factors include exchanging money with the public and working with volatile, unstable people. Working alone or in isolated areas may also contribute to the potential for violence. Providing services and care, and working where alcohol is served may also impact the likelihood of violence. Additionally, time of day and location of work, such as working late at night or in areas with high crime rates, are also risk factors that should be considered when addressing issues of workplace violence. Among those with higher risk are workers who exchange money with the public, delivery drivers, healthcare professionals, public service workers, customer service agents, law enforcement personnel, and those who work alone or in small groups.

How Can Workplace Violence Hazards Be Reduced?

In most workplaces where risk factors can be identified, the risk of assault can be prevented or minimized if employers take appropriate precautions. One of the best protections employers can offer their workers is to establish a zero-tolerance policy toward workplace violence. This policy should cover all workers, patients, clients, visitors, contractors, and anyone else who may come in contact with company personnel.

By assessing their work sites, employers can identify methods for reducing the likelihood of incidents occurring. OSHA believes that a well-written and implemented workplace violence prevention program, combined with engineering controls, administrative controls, and training, can reduce the incidence of workplace violence in both the private sector and federal workplaces.

This can be a separate workplace violence prevention program or incorporated into an injury and illness prevention program, employee handbook, or manual of standard operating procedures. It is critical to ensure that all workers know the policy and understand that all claims of workplace violence will be investigated and remedied promptly. In addition, OSHA encourages employers to develop additional methods as necessary to protect employees in high-risk industries.

Investigating Workplace Violence Complaints

Security managers should take violence complaints seriously. Employees should be encouraged to report potential volatile situations to security. The security investigator should document the complaint and follow up with local law enforcement if necessary. Depending on the act or circumstances, OSHA may order an inspection and investigation of the incident. Witnesses should be interviewed and statements recorded and signed for potential legal action, as well as documenting trends or the need for additional security.

References

Deborah Lamm Weisel, Graffiti, Problem-Oriented Guides for Police Series No. 9, U.S. Department of Justice, 2002.

Enforcement Procedures for Investigating or Inspecting Workplace Violence Incidents, OSHA Directive CPL 02-01-052, September 8, 2011.

Graffitihurts.org.

Guidelines for Preventing Workplace Violence for Health Care and Social Service Workers, OSHA Publication 3148-01R, 2004.

Jonathan White, *Terrorism and Homeland Security*, Belmont, CA: Thompson Wadsworth, 2008.

National Gang Intelligence Center, *National Gang Threat Assessment: Emerging Trends*, GPO, Washington, DC, 2011.

OSHA.gov.

Preventing Violence against Taxi and For-Hire Drivers, OSHA Fact Sheet, April 2010.

Recommendations for Workplace Violence Prevention Programs in Late-Night Retail Establishments, OSHA Publication 3153-12R, 2009.

4

Risk Assessment and Security Surveys

Tom Woodall and Bobby Ricks

Contents

Assessing and Managing Risk

Risk is the possible outcome that may cause a negative impact on an organization, impeding its ability to accomplish its mission or goals. A negative impact can be as simple as the theft of materials or as complex as a scheme involving industrial espionage. There is a difference between risk assessment and managing risk; risk assessments are conducted by security professionals, and risk is managed by senior management. It is the job of the security professional to evaluate all potential risks.

Risk assessments evaluate potential impact from man-made or natural acts, ranging from crimes against persons or property to earthquakes or tornados. Risk assessment is a team effort. The end result is better with more people participating in the process: building management, law enforcement, real estate personnel, and emergency management specialists, to list a few. Getting a diverse outlook at the risk will generate better ways to decrease the risk.

Before you begin the risk assessment process, you must first determine management's risk tolerance—acceptance of the risk outcome or taking action to prevent or mitigate the risk. A nuclear power plant cannot afford to take security risks because of the impact should a security breech occur. Other industries assume the risk and pass the cost on to their customers. Retail companies have a set amount of loss that is acceptable before they take action, like one retailer who determined a 3% loss from theft is acceptable. The risk tolerance in a company or organization will be affected by the security knowledge and backgrounds of senior management. In most cases, the senior managers will have little to no security experience. It is the security professional's job to explain the risk assessment and justify any countermeasures that are recommended to mitigate risk in terms that the CEO/CFO will understand.

The first step is to list all possible risks and determine the probabilities of those risks occurring. It is not possible to eliminate all risk, but you mitigate the risk or reduce the impact of an event. It is extremely important that you consider the potential impact on company operations when making recommendations to mitigate risk. Define each risk as probable or possible. Possible risks are events that could occur but are not likely. A probable risk meets these criteria but also is likely to occur at a higher rate. A security manager for an airline company may fear terrorists firing a missile at an airplane. This risk is categorized as possible instead of probable. A more probable risk is a bomb placed on an airplane. Operational managers are more likely to spend money to mitigate risks you can show as probable. Operational managers usually have little physical security experience. Write your evaluation of potential risks and recommendations in terms management will understand.

Consider three main factors when doing a risk assessment:

1. **Identify someone or some action that will cause a negative effect.** When evaluating this threat, you must determine that it has an intent to do harm, look at the history of the group for doing harm, and look at its capability to take action. Identifying all possible factors is very important to this process. Making sure that all factors are valid is even more important.
2. **A negative impact if the action is accomplished (against person, property, or information).** When doing this, you need to consider not only the impact on the location you

are assessing, but also the potential impact on others. For example, an assessment of a weapons manufacturing facility revealed it was the only facility that manufactured shells for the military. The negative impact to this site would be critical should a hazardous event occur.

3. **A vulnerability to the existing security that would allow the act to be accomplished**. This could be poor or improper hardware or a policy or procedural gap or loophole. One of the most common vulnerabilities to hardware is the improper installation or operation of systems.

The end product will be a list of recommendations to mitigate the identified risk. Multiple recommendations for each risk will allow executive management to assess the level and need for the protection of assets. Each recommendation should contain how much the risk would be mitigated, the cost of the measure, and potential impact should the recommendation be ignored.

There are many factors management must consider when deciding to spend money on physical security. Once you have finished your risk assessment and provided it to management, your job is done. Managers can do something or do nothing. The potential result is that something happens or nothing happens. Doing nothing may create a liability. The New York Supreme Court ruled that the New York Port Authority was over two-thirds liable for the bombing in 1993 since it chose to ignore an existing vulnerability. The Port Authority chose to continue allowing public parking in its basement after it became public knowledge that doing so was a security risk (Figure 4.1).

	DO SOMETHING	DO NOTHING
Something happens	You were prepared. Evaluate response and make changes to improve.	Very bad.
Nothing happens	You are ready. Maintain level of readiness.	Good, but is this by chance or based on a reasonable assessment of probabilities?

Figure 4.1 Action rubric.

Risk Management

Once a risk has been determined by the security professional, one of five actions will be taken by the operational manager. At this point the liability for this risk is accepted by the manager and the role of the security professional is finished.

1. Accept the risk. It is determined that the risk is at an acceptable level and no further action will be taken. This is also known as the cost of doing business. In some cases there is nothing that can be done to mitigate the risk and business has to continue.
2. Avoidance. Management can remove threatened assets to eliminate the risk. For example, when storms threaten military bases in south Florida, they move their planes and ships out of the area.
3. Reduction. The number of assets on site can be reduced to minimize the impact of a loss. For example, convenience stores use cash drops to reduce the money in a register. In the event of a robbery, the loss has been minimized to the amount in the cash drawer for operations.
4. Spreading. This is where you harden the target to increase the risk to the bad guys. The risk becomes greater than the reward.
5. Transfer. The most common transfer is through risk insurance. The risk is insured to cover the loss. Some of the risk is accepted by the level of deductibles on insurance policies.

A Risk Assessment Matrix

In the 1980s, the General Services Administration created a risk assessment matrix to identify and prioritize the security risks to a facility. A team of law enforcement personnel, life safety specialists, facility managers, engineers, and real estate specialists designed the program. The risk assessment matrix (RAM) consisted of a series of questions with points assigned to each answer so each facility is evaluated on the same criteria. The RAM had 30 main questions with an additional 18 subquestions. The questions included current and past crime statistics, facility attributes such as construction and types of security hardware, and information regarding the crime statistics in the area of the facility. The higher the score, the more risk was associated with the facility. Security resources were allocated to mitigate risks in priority order. Security specialists conducting the risk assessment would have to justify continuing, upgrading, or discontinuing security measures not recommended by RAM.

Decisions matrices are used throughout industry to identify needs versus wants and prioritize project management. An example of a decision matrix is in Figure 4.2.

In the matrix, decision criteria are listed and ranked on the top row. Options are listed on the left column, each being evaluated by the criteria and given a numerical value. With the criteria ranked and

Decision Matrix

	CRITERIA/ WEIGHT	CRITERIA/ WEIGHT	CRITERIA/ WEIGHT	CRITERIA/ WEIGHT	CRITERIA/ WEIGHT	
						Total
Options	1	2	3	4	5	
1						
2						
3						

Figure 4.2 Decision matrix.

	CRITERIA/WEIGHT	CRITERIA/ WEIGHT	CRITERIA/ WEIGHT	
	Meets statutory requirement, 41 USC § 421	Best practice dictates lock withstand weather extremes	Cost	Total
Options	3	2	1	
1 Greatlock 2050	Meets standard with a rating of 4/5 12 (score of 4 times rank of 3 = 12)	Withstands weather extreme to 0°F 3/5 6 (score of 3 times rank of 2 = 6)	Lowest price among choices 5/5 5 (score of 5 times rank of 1 = 5)	23
2 Fantastic Lock 500	Meets standard with a rating of 5/5 15	Withstands weather extreme to −20°F 4/5 8	Highest price among choices 1/5 1	24

Figure 4.3 Decision matrix.

evaluated, the matrix reveals the option that best fits the requirements. Criteria can be valued equally or weighted by the priority of the criteria. For example, with two criteria, one being a statutory requirement and the other being good practice, the statutory requirement may be given greater weight. Your scoring would give more weight to the statutory criteria over the good practice criteria (see Figure 4.3).

In Figure 4.3, the example is for a lock that must meet three criteria: statutory, best practices, and cost. Each criterion is weighted, and the corresponding value is multiplied by the weight. According to the matrix, the Fantastic Lock 500 is the best choice according to the criteria.

The Survey

A security survey tells us where we are, where we need to be, and how to get there. The survey identifies all requirements associated with the application of physical security hardware and policies to counter the threats. The security survey consists of gathering data necessary to do a thorough analysis of risk so appropriate recommendations can be made. Where we are is the current facility security measures on hardware and policy. Where we need to be is the level of security to protect organizational assets. How to get there is assessing the current security measures' effectiveness against foreseen threats and the level of risk the organization is willing to accept.

A survey may be required by statutory or regulatory requirements. For example, the U.S. Congress passed a law authorizing the protection of government property. The law authorized an administrative bureau to establish regulations on physical security. Insurance companies may require a physical security survey before insuring a facility or asset. Contractual agreements with other companies may require a level of security to protect assets owned by them in your possession. A legal duty to protect persons on the site would benefit from a survey to assure reasonable threats have been identified and addressed in the current security program. Surveys should be conducted when considering a site, change in facility mission, change in neighborhood, or when intelligence indicates an increased threat.

The survey team may consist of one person to a team of specialists, each examining their specialty and making recommendations to the team leader. Surveyors may be in-house, equipment suppliers, consultants, and systems integrators.

A physical security survey begins with presurvey planning. Previous security surveys, if available, should be reviewed. Obtain a map of the site showing buildings, utilities, and roads around the site. Determine if there are changes not noted on blueprints or site plans. Identify future changes that may impact security, such as new construction or the modification of existing facilities that may impact a security upgrade. Identify the number of employees, and pedestrian and vehicle traffic patterns, e.g., loading dock hours, shift change. Copies of current security policies and security reports should be obtained and reviewed. Identify site restrictions such as security clearances, escort requirements, and on-site response force procedures.

Research the location. Identify the local neighborhood and crime patterns in the immediate area. Obtain crime statistics from the local police to determine possible threats or concerns for security and safety of personnel. Check with government planners for future development and zoning laws, restrictions, and ordinances that may affect security. Get copies of relevant fire and safety regulations.

Identify the assets on site, the location of the asset, and the cost of loss, compromise, and ease of replacement. The value of a specific item that is stolen or destroyed may have a severe economic impact, not only by the loss of the item, but also by the ability of the item to create revenue. An oven for a restaurant may be easily replaced, or it may take several days to get a commercial oven delivered, and the restaurant can't operate until a new one is purchased and installed. The "secret" recipes relied on by the restaurant can bring everything to a halt in the kitchen, as well as the threat of competitors getting your trade secret.

Review the present security hardware and policies to determine if they are sufficient to deter or prevent the loss or compromise of the key assets. Threat decomposition looks at each potential threat and the ability of that threat to defeat the security measures in place. For example, a truck yard housing trailers loaded with your product is susceptible to theft and vandalism. Your threat decomposition would include identifying potential people who may be a threat (auto thief or gang) and their ability to accomplish their threat (theft or vandalism).

Site Visit and Adjoining Property—Neighborhood

With this information, visit the site for a physical assessment of the site, adjoining property, and the area surrounding the site. The site visit is to validate the presurvey data and allow for a physical inspection of the facilities and the surrounding area. The survey begins with a tour for a general assessment of the site, appearance, and attitudes toward security. The grounds and specific buildings are examined in detail. What security policies are in place? Is there an equipment and inventory control policy? What is the appearance and security of parking areas? How are visitors controlled? How are the grounds protected? What is the construction of the buildings? What type of security hardware is on the buildings? How is access controlled?

Is lighting adequate for safety and security? Are windows and other openings secured? What interior security measures are in place? Does the current security equipment work? Is it used? Are security policies followed? Is there an on-site response force? What is the response time for local police, fire, EMS? What are the function and duties of the on-site response force?

The survey will identify which structures are in need of additional or updated security hardware and policies. The site visit can identify physical or environmental factors present that might affect the operation of selected security devices. For example, a perimeter check would disclose a hilly terrain making line-of-sight equipment, such as microwave transmission, less effective. Recommendations will include the tentative physical location for additional devices as well as possible structural modifications and upgrades necessary to support the additional physical security equipment.

The surrounding area should be surveyed for proximity to potential threats: terrorist targets, hazardous materials facilities, shelters, or jails/prisons. A federal facility in a major U.S. city was located between two homeless meal kitchens. The facility experienced thefts and vandalism in its parking areas higher than normal for the area as people moved from one kitchen to another.

When the survey is completed, a survey report is written to document the current security posture and recommendations to management. Findings should be straightforward with recommendations based on the threat and need. Recommendations should be based on the ability to counter the threat, not on budgetary or other concerns. Management may choose not to implement a recommendation based on budget or other concerns, but the role of the surveyor is to report the facts and make recommendations for appropriate countermeasures.

List recommendations in priority order of requirement and need. For example, statutory and regulatory requirements would be high priority because of the risk of civil or criminal sanctions for failure to provide a level of security, as well as the potential for risk to assets for inadequate security and potential lawsuits for breach of duty. When determining the cost of a countermeasure, include all associated costs, such as maintenance or monitoring, for example, recommending $300 to install a security light, as well as considering the cost of a pole or running electricity to the light. Common shortfalls are maintenance costs for systems, underestimating the cost of installation for a system, and forgetting to include the cost of power to operate equipment.

Some recommendations may cost little to implement. For example, if there is a failure to follow established security protocols, the solution may be as simple and inexpensive as enforcing the policy. The 80/20 rule is a hypothesis that you can solve 80% of your problems by addressing the top 20% of your deficiencies. While focusing on your top priorities, solutions that cost nothing to implement and problems that can be resolved under routine costs and maintenance should be accomplished immediately.

5

Community Surveys

JARRED BALL

Contents

Being a successful security manager involves more than just understanding your facility's security issues. It is realizing how the larger community around you affects your facility and the role your facility plays in the overall community's security and functionality. Everything in the community is interdependent. For example, a factory is dependent on the power grid, water system, and transportation infrastructure of the surrounding community. If one or more of these sectors are destroyed or disrupted, the factory will either stop operations or have them severely impacted. Community risk assessment is a critical part of an overall security/mitigation plan for any security manager. Community risk assessments are composed of several variables, but they can be summed up into five main areas: identifying hazards, identifying critical infrastructure, realizing vulnerabilities to hazards/corrective actions, quantifying risk, and developing a community vulnerability assessment team. As you go through this chapter, you will examine in detail the five main areas of a successful community vulnerability assessment program and how to implement the plan to your facility. You also will gain an understanding of the interdependencies of the local community and how various threats can expose fragile areas of the system.

Identifying Hazards

The first area of the community risk assessment process is identifying hazards. Hazard is defined as exposure to danger. Hazards are everywhere, no matter where you live or work. You can never eliminate hazards completely, but they can be mitigated and lowered to acceptable levels in both the community and facility. To begin to lower and mitigate effects of hazards in our community and facility, the possible hazards must be identified.

A tool that can assist in the process is the all-hazard assessment. It is an inventory of all hazards that have happened within the local community in the past. The assessment will give an understanding of the potential risks the community and facility may face in the future. As you inventory the hazards, be sure to document them with research.

All hazards that have occurred should be included in the assessment, even if they have only occurred once. Examples of hazards to research include major disasters such as tornadoes, fires, and floods, manufacturing accidents, and criminal acts, including bombings or shootings.

To conduct research on past hazards, several areas can be explored. The first is local newspapers, many of which maintain extensive archives of major events. Local museums are another good source of information on local disasters that have occurred. Libraries are also a good place to conduct research. And finally, Internet searches for disasters in a particular area can turn up events that may have occurred 100 years ago that the people of the area had never heard of before.

The reason for documenting and researching all hazards that have occurred in an area is to give the security manager an understanding of all potential risks he or she may face. If it occurred once, it could occur again and should be planned for accordingly.

Next, it is important to understand the different types of hazards. This can help security managers get a handle on the multitude of risks that the facility and community may face. Hazards can be categorized in three different areas: natural, man-made, and accidental. First let's take a look at natural hazards.

Natural hazards are unexpected or uncontrollable natural events of unusual magnitude that threaten the safety/security of the community. This type of event usually causes widespread destruction of property or causes injury or death. Natural hazards include events such as hurricanes, earthquakes, tornadoes, floods, forest fires, and ice storms, to name a few.

Natural hazards cannot be prevented. Their effects, however, can be mitigated. This is where preplanning comes into play, which will be discussed further in the chapter. Natural hazards are usually widespread, which means that they will affect larger geographic areas. For example, a hurricane or earthquake may affect several cities and states.

For security managers, it is imperative to understand that a natural hazard does not have to hit your particular facility or immediate community to cause negative impacts at your site. A natural hazard may hit a couple of towns away and create problems on the utilities and transportation infrastructure that your facility depends on.

Man-made hazards are intentionally caused by man and can directly or indirectly cause severe threats to public health or well-being. Because their occurrence is unpredictable, human-caused disasters pose an especially challenging threat that must be dealt with through vigilance, preparedness, and response.

Man-made hazards include acts of terrorism, sabotage, and impropriety theft. Some examples of acts of terrorism would be bombings, mass shootings, assassinations, and hostage taking. Acts of terrorism could be carried out by individuals or groups. It is important to understand what terrorist groups operate in your community and region. This information can be found from several different sources, including local police reports, state/federal homeland security agencies, and private intelligence companies.

Sabotage is a deliberate action aimed at weakening an entity through subversion, obstruction, disruption, or destruction. Sabotage can be carried out by outside individuals or from disgruntled employees inside the facility. An example of sabotage is destroying equipment or processes in a facility with a physical or cyber-attack.

Impropriety theft is the stealing of information from a company and distributing it to a competitor. This mainly occurs in the private sector, but can happen in government, especially when dealing with national security information, and is usually conducted by internal employees with access to the information. Impropriety theft can destroy a company financially by losing vital business information to a competitor.

After researching and compiling a list of all known hazards in your community, the next step will be to conduct a hazard probability assessment, which categorizes hazards and assigns values on the basis of three factors: existence of hazard, history of hazard, and future threat potential. This tool will allow the security manager to apply some numerical values to the hazards that were documented in the all-hazard assessment and give some focus on what is the most likely hazard to occur in a given area.

Existence is the presence of a hazard within the jurisdiction, region, or state. Credible intelligence/information must exist in order to assign a rating for this factor; otherwise, it is given a value of zero.

History demonstrates past threat activity over a period of time, or a recorded violent criminal history. The hazard or event you are analyzing must have occurred once before in history, in your organization or in your location, in order to assign a rating to this factor; otherwise, it is given a value of zero.

Future threat potential is credible information indicative of preparations and capability for a specific threat or risk occurring in the future. In order for this factor to be assigned a value, evidence such as police reports and weather predictions must exist; otherwise, the factor is given a value of zero. Figure 5.1 is an example of a fictional town's hazard probability assessment. Numbers used for the hazard probability assessment will be subjective in nature.

Motown's Hazard Probability Assessment (HPA)				
Hazard	Existence (0–5)	History (0–5)	Future Threat Potential (0–5)	Probability Level
Terrorist	1	0	1	2
Tornado	1	1	2	4
Snowstorm	1	0	2	3
Hazardous Material Release	1	0	3	4
Fire	1	1	1	3

Figure 5.1 Motown's hazard probability assessment (HPA).

Identifying Critical Infrastructure

Now that the hazards have been identified and a hazard probability assessment is complete, we will examine what infrastructure in the community is critical to the everyday functions of society. Critical infrastructure is defined as those systems and assets, whether physical or virtual, so vital to the United States that the incapacity or destruction of such systems and assets would have a debilitating impact on security, national economic security, national public health or safety, or any combinations of those matters. According to Presidential Policy Directive/PPD-21, "Critical Infrastructure and Resilience," there are 16 designated critical infrastructure sectors:

1. Chemical
2. Commercial facilities
3. Communications
4. Critical manufacturing
5. Dams
6. Defense industrial base
7. Emergency services
8. Energy
9. Financial services
10. Food and agriculture
11. Government facilities
12. Healthcare and public health
13. Information technology
14. Nuclear reactors, materials, and waste
15. Transportation systems
16. Water and wastewater systems

Key resources are individual targets whose destruction would not endanger vital systems but could create local disaster or profoundly damage the nation's morale or confidence. An example of a key resource would be the Liberty Bell or other national monuments. It is crucial to understand what is critical to your community and facility as the security manager, because any disruption of these systems can have an adverse effect on your facility's normal functions.

After examining the various sectors of critical infrastructure/ key resources, the security manager should begin compiling a list of the critical infrastructure/key resources that his or her facility is dependent upon for normal everyday functions. This should include locations of the sites as well. The next step would be to preplan how the facility would function and for how long if one or more critical infrastructure/key resources were compromised by a hazard. For example, how long could the facility function if the community's water supply or electrical grid were damaged or destroyed? Are there backup water and power supplies to keep operations at normal for a period of time?

Vulnerabilities to Hazards

Realizing vulnerabilities to hazards is the next key to the community vulnerability assessment process. Vulnerability means exposure to risk, such as living by a river that floods its banks on a regular basis. Vulnerabilities are everywhere. Some are very evident and plain to see while others are not so clearly visible. Once vulnerabilities to hazards are understood, then the process of mitigating the effects of hazards can begin.

First, let us examine the common vulnerabilities to facilities by hazard type. We will begin with natural hazards. Earlier we looked at some different examples of natural hazards. When looking at your community and facility, ask yourself these questions: Is the facility located in a hurricane-prone area such as a coastline? Is the facility in or near Tornado Alley or close to a major fault line?

The first area of vulnerability is the building itself. What is the facility constructed of? Can the facility withstand hurricane force winds? Are there special brackets or supports to handle a certain magnitude of earthquake? Can water be pumped out of the facility if flooding were to occur? Some of these vulnerabilities can be expensive to correct, but can save money in the long term and, more importantly, save lives of the employees.

Some other areas to consider when dealing with exterior vulnerabilities and structural deficiencies are:

- Soil composition
- Elevation

- Proximity to bodies of water such as lakes and rivers
- Construction materials used on the facility, including roof, windows, and exterior walls
- Backup power and water generation capabilities

The vulnerabilities to natural hazards can be mitigated to both facilities and the community at large by using different construction materials and installing backup power and water generation systems.

Man-made hazards include events like bombings, shootings, and sabotage. When discussing bombing attacks, does the facility have blast-resistant materials on the doors and windows to prevent glass from shattering and becoming projectiles? What about bollards or other barriers to prevent potential vehicle-borne improvised explosive devices (VBIEDs) from gaining close proximity to the facility? These are some common vulnerabilities when looking at explosive-related man-made hazards.

Shootings involve issues with access control, which means how people can enter the facility in a secure way. Access control can be instituted by using a number of existing technologies. A common technology is the proximity card reading system. This system uses picture identification cards with embedded microchips in them that will electronically open facility doors when swiped or held near a card reader. Using the software, individuals' entry can be limited.

Having on-site, armed security guards can be a very effective way to deter a shooter. They can respond quicker than waiting for local law enforcement and can eliminate or delay the threat. Having security cameras can be a deterrent but will not stop a shooter. Implementing policies and procedures for shooting incidents can be a cheap and effective way to mitigate the vulnerabilities of shooting incidents. Practicing lockdown drills and evacuations on a regular basis is a great way to increase the awareness and education to the threat of shooting incidents.

Vulnerabilities to sabotage can be addressed with policy and procedures, to include preemployment background checks. Security cameras can also deter sabotage along with on-site security guards. Installing fencing, exterior lighting, and intrusion alarm systems can also work to mitigate vulnerabilities to sabotage. Internet security solutions such as firewalls, password protection/encryption, and routine employee background checks can limit the possibility of electronic sabotage or theft of proprietary information from the facility.

Consequences of accidental hazards can be just as damaging and costly to the facility and community as any natural or intentional man-made hazards can be. Vulnerability to accidental hazards can be mitigated by employee training, policy and procedures, and installation of proper equipment to have a safe working environment. Vulnerability to accidental hazards can sometimes be a regulatory issue with agencies such as the Occupational Safety and Health Administration (OSHA). It is important to check with the proper regulatory entity to make sure the facility is in compliance with all safety regulations.

Quantifying Risk

Quantifying risk is the next step in the community vulnerability assessment process. Risk is defined as the probability that an event will occur and the consequences of its occurrence. There are four steps in quantifying risk according to the Federal Emergency Management Agency (FEMA): (1) determine asset value, (2) determine threat value ratings, (3) determine vulnerability rating value, and (4) determine relative risk for each threat against each asset.

These numbers will be subjective in nature. It is impossible to take away all subjectivity from rating risks and threats for possible future events. Each one of the above steps will have a range of numbers, for example, 0–5, with 0 being low and 5 being high.

It is the best educated guess based on history, current information, and intelligence and vulnerability against the threat times the consequences if the threat did occur. Quantifying risk can be complex; however, keep everything in perspective. Use number values that are consistent and fit your community or facility.

A common, simple formula for quantifying risk is

$$\text{Risk} = \text{Threat} \times \text{Vulnerability} \times \text{Consequence}$$

Figures 5.2 to 5.5 are examples of FEMA's quantifying risk tables.

Very high	10
High	8–9
Medium-high	7
Medium	5–6
Medium-low	4
Low	2–3
Very low	1

Figure 5.2 Risk factor definitions.

	LOW RISK	MEDIUM RISK	HIGH RISK
Risk factors total	1–60	61–175	≥176

Figure 5.3 Total risk color code.

FUNCTION	CYBER ATTACK	ARMED ATTACK (SINGLE GUNMAN)	VEHICLE BOMB	CBR ATTACK
Administration	280	140	135	90
Asset value	5	5	5	5
Threat rating	8	4	3	2
Vulnerability rating	7	7	9	9
Engineering	128	160	384	144
Asset value	8	8	8	8
Threat rating	8	5	6	2
Vulnerability rating	2	4	8	9

Figure 5.4 Critical functions.

FUNCTION	CYBER ATTACK	ARMED ATTACK (SINGLE GUNMAN)	VEHICLE BOMB	CBR ATTACK
Site	48	80	108	72
Asset value	4	4	4	4
Threat rating	4	4	3	2
Vulnerability rating	3	5	9	9
Structural systems	24	32	240	16
Asset value	8	8	8	8
Threat rating	3	4	3	2
Vulnerability rating	2	4	8	9

Figure 5.5 Critical infrastructure.

Developing a Community Vulnerability Assessment Team

Once you have worked through the steps of identifying hazards, identifying critical infrastructure, realizing vulnerabilities to hazards/corrective actions, and quantifying risk, it is time to conduct a risk vulnerability assessment (RVA) for your facility and community. The first step in conducting an RVA is to construct an RVA team.

When selecting members for an RVA team it is important to incorporate individuals from different disciplines. This holistic approach will ensure that the community and facilities that are assessed will have a multidisciplinary view of their vulnerabilities from different threats. Some disciplines to consider with team selection include police, fire, EMS, emergency management, military, bomb squad/explosive ordinance, engineers, and tactical team members.

The problem with having a team of all fire or police personnel is that they tend to look at the problem and offer solutions with one set of training as background. In doing so, they may miss a more glaring vulnerability that another discipline might expose.

When selecting team members, it is important that they are experienced in their field of work and have some homeland security training, if possible. Confidentiality is another very important issue to discuss with the team. Vulnerabilities from threats that communities and facilities may have should be kept with utmost confidentiality. If these weaknesses were to get in to the wrong hands, it could endanger the community by placing those sites at risk from an adversary attack. Also, it could give an advantage to private companies' competitors by allowing them to exploit those vulnerabilities for their own financial gain.

The RVA team should institute operational security (OPSEC) measures, which is a military-based process for dealing with confidential information that may be security issue or proprietary business processes. Some ways to keep information secure include:

- Shred sensitive documents no longer in use.
- Keep sensitive materials under lock and key.
- Password protect all computers and electronic data.
- Do not disseminate sensitive information to the media.
- Do not discuss sensitive information on social media platforms.
- Conduct background checks on individuals who have access to sensitive information. (including janitorial/contract employees who are unescorted).
- Clean all cameras and computers when finished from working out in the field.
- Place "sensitive information" or some other form of disclaimer on the header of sensitive documents.

Conducting the Assessment

Now that the team is assembled, it is time for the assessment. When assessing a facility for either natural or man-made disasters, the first step is to examine the exterior facility and grounds. When conducting a walk around the exterior facility and grounds, the assessor should take pictures of the building and the surrounding campus. It is important to make detailed notes of what the facility is constructed of and exterior entrances. The assessor should also document any fences, exterior lighting, and security guards.

After a thorough tour of the exterior of the facility, the next step is the facility interior. When examining the interior portions of the facility, an important thing to look at is the overall layout of the interior. It is helpful to try to obtain copies of the blueprints of the facility to keep in the team's file. Maintenance workers or janitors are a very valuable source of information about the facility. If permissible, interview a couple of employees to help the team characterize the facility and the nature of its daily operations.

If possible, try to obtain copies of the emergency operations procedure handbook and any other safety policy materials. By examining these documents, the team can understand the preparedness levels of the facility and its personnel. Sometimes vulnerabilities can be solved by simply adding a policy or procedure.

Facilities operations can be completely different from daylight hours to overnight. The team should try to come back to the facility at night to take a quick inventory of the overall security climate at dark. Also, some facilities are busier during different times of the year. These are important factors to consider when conducting a comprehensive assessment. Make sure to document the power and water systems of the facility, including the HVAC systems. Discuss backup plans in case loss of water or power occurs at the facility.

Study the facility's access control system, key control, and what, if any, background checks are performed on individuals before they are hired. And finally, examine the facility's Internet and computer systems. Do they back up their data? And if so, do they back it up off-site? What kind of information security do they employ? Do they have antivirus software or an Internet firewall? Do they use encryption technology?

Never show up to a facility unannounced! Walk-throughs should be scheduled in advance, and the facility should have a representative escort the assessors at all times. Before the assessment begins, the RVA team should hold an orientation with facility management to discuss the purpose and schedule of the assessment.

Ask permission before taking any pictures. Some facilities may not want any pictures taken. After the assessment, the RVA team should hold an out briefing with the facility to discuss any pertinent issues that the team discovered during the assessment. Consider creating a professional report for the facility management after the team detailing the assessment and any vulnerabilities/recommendations they might employ to address the vulnerabilities. There are several vulnerability checklists available to assist the process to ensure that every area, exterior and interior, is covered during the assessment.

Risk and Vulnerability Assessments for Rural Communities. The Northwest Arkansas Community College is part of the Rural Domestic Preparedness Consortium (RDPC). www.ruraltraining.org

Building Design for Homeland Security, Unit 5, "Risk Assessment/Risk Management." The tables are dealing with quantifying risk. FEMA.gov

Understanding, Assessing and Responding to Terrorism, 2007, Benett T. Brian. Hoboken, NJ: John Wiley & Sons.

Threats to Homeland Security, 2008, Kilroy J. Richard Jr. Hoboken, NJ: John Wiley & Sons.

6
OSHA

THOMAS D. SCHNEID

Contents

For security professionals with safety and health responsibilities, a knock on the door by an Occupational Safety and Health (OSHA) compliance officer can be a bewildering and dreaded experience. Who is OSHA? Why are they here? What is going to happen? However, this experience can be positive if the security professional has adopted a proactive approach and has created and maintained a safe and healthful work environment for his or her employees prior to the inspection and properly prepares for the "knock on the door." Security professionals with safety and health responsibilities should become knowledgeable concerning the safety and health standards (www.osha.gov) and requirements and make their safety and health responsibilities a priority within their daily activities.

The Federal Occupational Safety and Health Act of 1970

With the passage of the then controversial OSH Act in 1970,[*] federal and state governmental agencies became actively involved in managing health and safety in the private sector workplace.[†] Employers were placed on notice with the passage of the OSH Act that unsafe and unhealthful conditions and acts would no longer be permitted to endanger the health, and often the lives, of American workers. In the early years, employers were often forced, under penalty of law, to address safety and health issues in their workplaces.

Today, the structure and activities developed under the OSH Act are virtually unchanged since this law was enacted. The basic methods for enforcement, standards development, and promulgation, as well as adjudication, remain intact. OSHA has, however, added many new standards in the past ±40 years based primarily on the research conducted by the National Institute for Occupational Safety and Health (NIOSH) and recommendations from labor and industry. In addition, OSHA has revisited several of the original standards in order to update or modify the particular standard. The Occupational Safety and Health Review Commission (OSHRC) and the courts have been very active in resolving many disputed issues and clarifying the law as it stands.

Legislative History

Throughout the history of the United States, the potential for the American worker to be injured or killed on the job has been a brutal reality. Many disasters, such as that at Gauley Bridge, West Virginia,[‡] fueled the call for laws and regulations to protect the American worker. As early as the 1920s, many states recognized the safety and health needs of the industrial worker and began to enact worker's compensation and industrial safety laws. The first significant federal legislation

[*] 29 USC Section 63 et. seq.

[†] Note: Many public sector employers have adopted OSHA jurisdiction over their workplaces.

[‡] Page, J., and M. O'Brien, *Bitter Wages* (1973). During the construction of a tunnel in 1930–1931, 476 workers died, and approximately 1,500 were disabled, primarily by silicosis. New York: Grossman Publishers.

was the Walsh-Healey Public Contracts Act of 1936, which limited working hours and the use of child and convict labor. This law also required that contracts entered into by any federal agency for over $10,000 contain the stipulation that the contractor would not permit conditions that were unsanitary, hazardous, or dangerous to employees' health or safety.

In the 1940s, the federally enacted Labor Management Relations Act (Taft-Hartley Act) provided workers with the right to walk off a job if it was "abnormally dangerous."* Additionally, in 1947, President Harry S Truman created the first Presidential Conference on Industrial Safety.

In the 1950s and 1960s, the federal government continued to enact specialized safety and health laws to address particular circumstances. The Coal Mine Safety Act of 1952, the Maritime Safety Act, the McNamara-O'Hara Public Service Contract Act (protecting employees of contractors performing maintenance work for federal agencies), and the National Foundation on the Arts and Humanities Act (requiring recipients of federal grants to maintain safe and healthful working conditions) were all passed during this time.

The federal government's first significant step in developing coverage for workplace safety and health was the passage of the Metal and Nonmetallic Mine Safety Act of 1966. Following the passage of this act, President Lyndon B. Johnson, in 1968, called for the first comprehensive occupational safety and health program as part of his Great Society program. Although this proposed plan never made it to a vote in Congress, the seed was planted for future legislation.

Coverage and Jurisdiction of the OSH Act

The OSH Act covers virtually every American workplace that employs one or more employees and engages in a business that affects interstate commerce in any way.† The OSH Act covers employment in every state, the District of Columbia, Puerto Rico, Guam, the Virgin Islands, American Samoa, and the Trust

* 29 USC Sections 401–531.

† 29 CFR Section 1975.3(d).

Territory of the Pacific Islands.* The OSH Act does not, however, cover employees in situations where other state or federal agencies have jurisdiction that requires the agencies to prescribe or enforce their own safety and health regulations.† Additionally, the OSH Act exempts residential owners who employ people for ordinary domestic tasks, such as cooking, cleaning, and child care.‡ It also does not cover federal,§ state, and local governments¶ or Native American reservations.**

The OSH Act requires that every employer engaged in interstate commerce furnish employees "a place of employment ... free from recognized hazards that are causing, or are likely to cause, death or serious harm."†† To help employers create and maintain safe working environments and to enforce laws and regulations that ensure safe and healthful work environments, Congress created OSHA, to be a new agency under the direction of the Department of Labor.

Today, OSHA is one of the most widely known and powerful enforcement agencies within the federal governmental structure. OSHA has been granted broad regulatory powers to promulgate regulations and standards, investigate and inspect, issue citations, and propose penalties for safety violations in the workplace.

The OSH Act also established an independent agency, the Occupational Safety and Health Review Commission (OSHRC), to review OSHA citations and decisions. The OSHRC is a quasi-judicial and independent administrative agency composed of three commissioners appointed by the president who serve staggered 6-year terms. The OSHRC has the power to issue orders; uphold, vacate, or modify OSHA citations and penalties; and direct other appropriate relief and penalties.

The educational arm of the OSH Act is the National Institute for Occupational Safety and Health (NIOSH), which was created as a

---

* Idem § 652(7).

† See, e.g., Atomic Energy Act of 1954, 42 USC § 2021.

‡ 29 CFR § 1975(6).

§ 29 USCA § 652(5) (no coverage under OSH Act when U.S. government acts as employer).

¶ Idem.

** See, e.g., Navajo Forest Prods. Indus., 8 OSH Cases 2694 (OSH rev. comm'n. 1980), aff'd., 692 F.2d 709, 10 OSH Cases 2159.

†† 29 USCA § 654(a)(1).

specialized educational agency of the existing National Institutes of Health. NIOSH conducts occupational safety and health research and develops criteria for new OSHA standards. NIOSH may conduct workplace inspections, issue subpoenas, and question employees and employers, but it does not have the power to issue citations or penalties. NIOSH is currently aligned with the Centers for Disease Control (CDC).

State Safety Plans

Notwithstanding OSH Act enforcement through the above-noted federal agencies, the OSH Act permits individual states to take responsibility for OSHA administration and enforcement within their respective boundaries. Each state possesses the ability to request and be granted the right to adopt state safety and health regulations and enforcement mechanisms. In Section 18(b), the OSH Act provides that any state "which, at any time, desires to assume responsibility for development and the enforcement therein of occupational safety and health standards relating to any … issue with respect to which a federal standard has been promulgated … shall submit a state plan for the development of such standards and their enforcement."* For a state plan to be placed into effect, the state must first develop and submit its proposed program to the Secretary of Labor for review and approval. The secretary must certify that the state plan's standards are "at least as effective" as the federal standards, and that the state will devote adequate resources to administering and enforcing its standards.†

In most state plans, the state agency has developed more stringent safety and health standards than OSHA, and has usually developed more stringent enforcement schemes.‡ The Secretary of Labor has no statutory authority to reject a state plan if the proposed standards or enforcement scheme are more strict than the OSHA standards, but can reject the state plan if the standards are below the minimum limits set under OSHA standards.§ These states are known as state plan

* Idem.

† Idem § 667(c).

‡ Some states adopted the federal OSHA standards as promulgated; however, other states enhance or modify the proposed standard to be more stringent.

§ 29 USC § 667.

states and territories. As of this writing, there were 21 states and 2 territories with approved and functional state plan programs.* Employers in state plan states and territories must comply with their state's regulations; federal OSHA plays virtually no role in direct enforcement. OSHA does, however, possess an approval and oversight role regarding state plan programs. OSHA must approve all state plan proposals prior to their enactment. It also maintains oversight authority to "pull the ticket" of any/all state plan programs at any time if they are not achieving the identified prerequisites.

OSHA Standards and the General Duty Clause

Promulgation of Standards

The OSH Act authorizes three ways to promulgate new standards: (1) national consensus standard, (2) informal (standard) rule making, and (3) emergency temporary standard. From 1970 to 1973, the Secretary of Labor was authorized in Section 6(a) of the act† to adopt national consensus standards and establish federal safety and health standards without following lengthy rule-making procedures. Many of the early OSHA standards were adapted from other areas of regulation, such as the National Electric Code and American National Standards Institute (ANSI) guidelines. However, this promulgation method is no longer in effect.

The usual method of issuing, modifying, or revoking a new or existing OSHA standard is described in Section 6(b) of the OSH Act and is known as informal rule making. This method requires providing notice to interested parties, through subscription in the *Federal Register*, of the proposed regulation and standard, and allowing parties the opportunity for comment in a nonadversarial, administrative hearing.‡ The proposed standard can also be advertised through magazine articles and other publications, thus informing interested

* The states and territories operating their state plan OSHA programs include Alaska, Arizona, California, Hawaii, Indiana, Iowa, Kentucky, Maryland, Michigan, Minnesota, Nevada, New Mexico, North Carolina (partial federal OSHA enforcement), Oregon, Puerto Rico, South Carolina, Tennessee, Utah, Vermont, Virginia, Virgin Islands, Washington, and Wyoming.

† 29 USC § 1910.

‡ 29 USC § 655(b).

parties of the proposed standard and regulation. This method differs from the requirements of most other administrative agencies that follow the Administrative Procedure Act* because the OSH Act provides interested persons the opportunity to request a public hearing with oral testimony. It also requires the Secretary of Labor to publish a notice of the time and place of such hearings in the *Federal Register*.

The most infrequently used method is the emergency temporary standard. The OSH Act permits, under Section 6(c), that the Secretary of Labor may immediately establish a standard if it is determined that employees are subject to grave danger from exposure to substances or agents known to be toxic or physically harmful, and that an emergency standard would protect the employees from the danger. An emergency temporary standard becomes effective upon publication in the *Federal Register*, and may remain in effect for 6 months. During this 6-month period, the secretary must adopt a new, permanent standard or abandon the emergency standard.†

Only the Secretary of Labor can establish new OSHA standards; however, recommendations or requests for an OSHA standard can come from any interested person or organization, including employees, employers, labor unions, environmental groups, and others.‡ When the secretary receives a petition to adopt a new standard or to modify or revoke an existing standard, he or she usually forwards the request to NIOSH and the National Advisory Committee on Occupational Safety and Health (NACOSH).§ Alternately, the secretary may use a private organization such as ANSI for advice and review.

The General Duty Clause

As stated above, the OSH Act requires that an employer maintain a place of employment free from recognized hazards that are causing, or are likely to cause, death or serious physical harm, even if there is no specific OSHA standard addressing the circumstances. Under Section 5(a)(1), the general duty clause, an employer may be cited for a violation

* 32 USC § 553.

† 29 USC § 655(c).

‡ www.osha.gov/OCIS/stand_dev.html

§ Ibid.

of the OSH Act if the condition causes harm or is likely to cause harm to employees, even if OSHA has not promulgated a standard specifically addressing the particular hazard. The general duty clause is a catchall standard encompassing all potential hazards that have not been specifically addressed in the OSHA standards. For example, if a company is cited for an ergonomic hazard and there is no ergonomic standard to apply, the hazard will be cited under the general duty clause.

OSHA Compliance Inspections

Reason for the Compliance Inspection

Security professionals should be aware that there is often a reason for the OSHA or state plan compliance inspection. Historically, the consistent number one reason why an operation is inspected is due to an employee complaint. For fatality or multiple injury situations, security professionals should be aware that there is a notification requirement for the company, or the security professional as the company representative, to contact OSHA. Other reasons that may generate an OSHA inspection include an industry-targeted inspection or a random selection. Security professionals should also become familiar with the OSHA initiatives, which include, but are not limited to, the Severe Violator Enforcement Program. Security professionals can find the most current information on the OSHA web page located at www.osha.gov.

OSHA Penalties

The enforcement of the OSH Act is primarily through monetary penalties; however, the OSH Act does provide for criminal sanctions for egregious situations. The monetary penalties range from *de minimus* or nonserious to serious, repeat, and willful and also possess categories for repeat violations, failure to abate violations, and failure to post the required documents. Violations of OSH Act standards or the general duty clause are categorized as de *minimis*, other (nonserious), serious, repeat, and willful. Monetary penalties assessed by the secretary vary according to the degree of the violation. Penalties range from minimal monetary penalty to 10 times the imposed penalty for repeat or willful violations.[*] Criminal sanctions are provided under the OSH Act. Additionally, the Secretary of Labor may refer willful violations to the U.S. Department of Justice for imposition of criminal sanctions. The penalty schedule identifying monetary penalty ranges and possible criminal sanctions can be found on the Occupational Safety and Health Administration website located at www.osha.gov

Citation, Appeal, and Defenses

In general, within 6 months of an inspection, OSHA will codify the identified alleged violations into a citation document and forward to the employer identifying the alleged violations, category of violation, and proposed monetary penalties. This document is usually sent to the employer via certified mail.

Security professionals must be aware that there is a 15-working-day limitation from the time the citation is received from OSHA following a compliance inspection. Security professionals should also be aware that failure to file the notice of contest in a timely manner can result in the loss of all appeal rights. Professionals should be aware that once the citation is issued, OSHA has the burden of proof to prove each and every alleged violation. The preponderance of the evidence standard is utilized in Occupational Safety and Health Review Commission hearings and rules of evidence are utilized throughout. In short, OSHA must prove each and every element of each and every alleged violation. In preparing the defenses, security professionals should be looking for deficiencies or lack of proof for each and every alleged violation.

Working with legal counsel, the first level of possible defenses usually involves the procedural aspects of the citation. These defenses are usually very technical in nature, such as a defect in the inspection procedure, and are usually not successful. OSHA inspectors are usually well educated in the procedural requirements. Two general defenses in the procedural area have been utilized, including the statute of limitation defense and the lack of reasonable promptness defense. The statute of limitation defense is often utilized when the citation is issued beyond 6 months from the time of the alleged violation. In general, if

[*] 29 CFR 1910.658(b).

the citation is not issued within 6 months from the date of the alleged violation, this is grounds to dismiss the citation. The lack of reasonable promptness defense, although not an absolute defense, generally involves a similar delay in issuing the citation, which prejudices the employer in preparing the defense. Other procedural defenses may be available depending on the circumstances.

Factual defenses are usually based on what the circumstances were at the time of the alleged violation, what was or wasn't observed by the compliance officer, what actions or inactions involved the employees, and related facts. Factual defenses are based primarily on the circumstances. One of the most often utilized factually based defenses where the organization possesses a well-established safety and health program is the defense of unpreventable employee misconduct. To prove this defense, the professional must document each and every aspect of the inspection and assist legal counsel in providing the established work rule or policy to prevent the violation from occurring; the work rule or policy was adequately communicated to all employees, including their supervisors or team leaders; the management team member took reasonable steps to discover the alleged violation; and management effectively enforced the work rule or policy that the employees violated.

Security professionals, working with legal counsel and the management team, should be able to identify applicable defenses depending on the facts and circumstances identified in the citation. Each and every detail in each and every alleged violation should be carefully scrutinized with each possible defense reviewed and analyzed. Diligence, creativity, and an eye to detail will permit the security professional and legal counsel to develop viable defenses to most alleged violations. If there are no defenses available, it may be time to attempt to negotiate reductions in the proposed categorization or monetary penalties through your good faith or other factors.

In summation, it is important for security professionals to acquire a working knowledge of the numerous safety and health standards and requirements as well as the OSHA enforcement process in order to safeguard the company, management, and workers. Security professionals with safety and health responsibilities should provide substantial efforts in creating and maintaining proactive safety and health programs that protect their employees from the various risks within the workplace. As with security, a proactive safety and health program can pay dividends in avoiding the potential of workplace injuries and illnesses.

Endnotes

i. 29 USC Section 658(b).

7

Fire Safety and Security

JAMES L. PHARR

Contents

Security managers may experience contact with the local fire department in a variety of situations, including emergency situations. Fire departments often respond to situations other than fires, including medical emergencies, rescue situations, and hazardous materials releases. In nonemergency situations contact is often focused on performing inspections to assure compliance with fire prevention codes. In other situations, fire services contact facilities to conduct pre-emergency planning activities that provide critical information that enhances their ability to act properly during an emergency response.

Fire prevention codes and standards are often cited in Occupational Safety and Health Administration (OSHA) regulations also. Knowing the genesis of these codes helps us understand how to apply them appropriately. Most often fire prevention codes and standards are developed by the National Fire Protection Association (NFPA).

The code writing process includes the ability for anyone to submit recommendations to a committee that drafts new and revised codes. Proposed language is promulgated to members of the NFPA for review and comments. The committee reviews comments and adjusts its suggestions as appropriate. The final draft is then submitted for a membership vote at the biannual conference where any member can speak for or against and vote on approval. If approved, the document is then reviewed by a committee that coordinates codes to prevent conflicting language between various codes and standards, and then the adopted code is published. Explanatory material is contained in appendices for sections marked with an asterisk in NFPA codes and standards.

Governments can adopt codes and standards as written, modify the language, or choose against adopting. Regardless of governmental action, these codes and standards are often central in litigation when incidents occur. Loss prevention managers should be aware of edicts contained, as they are often cited as the standard of care to protect people and property against fire.

Life Safety and Fire Prevention

Building codes, including structural, electrical, plumbing, and mechanical, specify how structures are constructed for stability and methods to provide utilities inside the building. When completed, rarely are additional inspections conducted under these codes; rather, application of fire prevention codes is normal. Fire prevention codes address maintenance of buildings as approved by construction codes, including use of materials and processes. The primary focus of all codes is life safety, assuring that regardless what happens to the structure, people are not injured. Multiple building codes are enforced throughout the United States; however, the preeminent code is the International Building Code.

Fire prevention inspections may identify situations where buildings are not being used as intended, and thus engage other disciplines to address concerns. Change of use is a frequent concern in the United States. This is largely due to changes in technology and repurposing buildings. Most fire services and building officials work closely with private concerns to find ways to permit desired use when consulted early in activities to repurpose structures and facilities—thus the recommendation to know your local fire officials. Many entities enforce the International Fire Code, which in turn references any of the codes and standards contained in the National Fire Codes promulgated by NFPA. NFPA 1, *Fire Code*, is promulgated by NFPA and outlines methods and procedures for providing basic fire safety in most structures and situations. However, this is not the only fire prevention code enforced by local and state governments.

Preemergency planning activities offer excellent opportunities to collaborate with fire services, including fire prevention personnel, to assure understanding of operations and expectations of the fire department. Building diagrams, sprinkler diagrams, hazardous materials lists, and hazardous locations are critical information for responders.

Unless the facility provides continual security services, methods to access facilities and buildings are recommended. Though fire departments theoretically could have keys required for access, logistics associated with having keys available to every location quickly are daunting if those keys are retained in emergency response vehicles. Many jurisdictions have high-security lock boxes that are accessible to both facilities and emergency responders. This facilitates access with a key designated for only that box, while responders hold a key that will open all such boxes in their jurisdiction. Facility personnel assure required keys and codes are inside the box, changing them when needed. During emergencies responders have immediate access to those keys. When access is required and keys are not provided, responders must force entry, which results in greater property damage than would be experienced when keys are readily available.

Facilities with hazards conditions or processes must communicate that information to response forces before emergencies. Understanding the general hazards of the processes, materials, or chemicals is critical to protecting people and properties. The Superfund Amendments and Reauthorization Act (SARA) of 1986 requires facilities to disclose general information of chemicals stored, used, and manufactured to local and sate governments. This allows those governments to develop plans to protect citizens should a release occur from the facility. Most state governments also have right-to-know laws that require providing similar information to responders who may come into the facility during emergencies. Facilities are not generally required to reveal

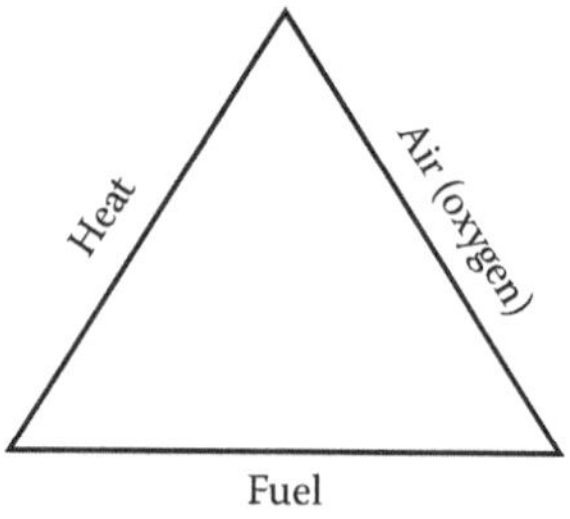

Figure 7.1 Fire triangle.

trade secret information, and most laws provide significant sanctions against public entitles that reveal trade secret information to those outside the company.

Fire prevention may be best understood by using the fire triangle to understand how fire works (Figure 7.1). In most cases three components are needed to have fire: heat, fuel, and oxygen.

Removal of one component of this fire triangle, or reducing that element below what is needed for fire, prevents fire from starting. Generally, air is present in all atmospheres; thus, fire prevention focuses on assuring fuel and heat are separated to prevent ignition. This is a simple concept that works in a high percentage of situations; however, some situations are outside the bounds. Greater examination of hazards is required in those rare situations.

Life Safety Concerns

Preventing life loss and serious personal injury is the primary concern for fire prevention and building codes. Many parts of each code are intended to assure fire and smoke conditions remain tenable for humans long enough for all occupants to get outside, and thus away from threats. Maintaining egress routes is one of the most important factors in assuring life safety.

Egress includes the exit access (path to the exit), exit (door), and exit discharge (path away from the exit). Egress routes must be clearly identifiable to occupants during emergency situations, such as low light and stressful conditions. Frequent exit drills familiarize occupants with multiple egress routes, and thus improve the ability to quickly identify preferred routes during emergencies.

Egress is designed on the width required to move humans through specific points at a rate sufficient to assure escape from the building in reasonable time. Required exit width varies with the number of occupants expected to move through the area and the physical location of that exit. Consult codes for guidance in determining required exit criteria.

Locked and blocked exit doors are frequently identified violations. Locked exit doors prevent rapid opening that is required during emergencies. Only one action is acceptable to facilitate opening exit doors. This action can be pushing a bar, turning a knob, or another easy-to-perform maneuver. Slide bolts are not acceptable on exit doors. Rarely are deadbolt locks acceptable, and never are they acceptable if a key is required on both sides. Generally, the best option is a panic bar or lever that opens the locking mechanism when depressed.

Aisles and hallways leading to exits must be clearly identifiable during emergencies. Illuminated exit signs, floor-mounted directional lighting, and other options are available. An oft-identified life safety violation is storage in corridors and aisles; however, these must remain unobstructed. Stairways are another area where storage is discovered on landings and beneath the lowest riser. Justifications include the stairwell is not used frequently and storage is placed to not be "in the way." Regardless, rarely is storage permitted in stairwells because full egress width is required and no combustibles may be in most of these life safety areas.

Once outside of structures, evacuees must be able to quickly and easily reach a location clear of the structure. Designated assembly locations where all occupants can be accounted for are essential for assuring all exited the building. Accounting of occupants reduces confusion and allows firefighters to focus on suppression rather than search activities.

Larger buildings, especially high-rise structures, may have areas of refuge rather than exit discharges. Areas of refuge are situated at various locations within the structure and are designed to afford increased protection for occupants until egress is possible. Some stairways feature smaller areas of refuge for occupants using wheel chairs or other mobility assistance devices. Those areas allow them to wait for rescue forces to arrive while remaining outside the egress pathway of others.

Fire wardens are usually assigned to assure that all employees and visitors have exited the building when evacuation is implemented. These persons must be trained prior to emergencies and should conduct a quick sweep of the area to assure all have left. The warden can assist those having difficulty moving to get out or call for help. Wardens must report their observations to a designated leader once they arrive outside.

Employee training and drills are essential to assuring egress during fires and other emergencies. All must recognize the fire alarm signal and immediately take action to exit the building. When critical processes must be shut down, those actions must be completed in proper order by specified procedure.

Occupancy-specific guidance is available in NFPA 101, *Life Safety Code*. This is a comprehensive collection of information regarding issues encountered in existing and new construction-based building use.

Housekeeping

Orderly arrangement within work areas provides the perception that safety, including fire safety, is taken seriously. Where order persists, more frequent checks usually result, and thus corrective actions are more likely. For example, where storage of papers and files is random and piles result, a failing electrical component may go undetected. Should that component fail result in overheating, ignition is more likely because additional fuel is present. Rapid flame development could result in hazard to employees. Reducing clutter can reduce chances fires will spread.

This concept transcends through manufacturing, storage, and all other areas where hazards are present. This is especially important to assuring egress routes are clearly identifiable during emergencies.

Flammable/Combustible Liquid and Gas Storage and Use

Ignitable liquids and gases pose significant fire and environmental hazards when used or produced in processes. Because these states of matter do not have a defined form, movement to areas where adverse results can occur is possible, if not likely. Liquid products may be visible, but vapors or gases are often undetectable without instruments.

The adage, "liquids don't burn, but the vapors emitted from liquids may," is factual.

Gases and liquids that will burn are classified by characteristics of vapor production, weight of vapors, and energy required for combustion to start. These characteristics include:

Gas: Materials that exist at standard temperature (68°F) and pressure (14.7 psi) in the gaseous state are known as gases.

Vapor: The gaseous state of materials that exist as liquids at standard temperature (68°F) and pressure (14.7 psi) is known as vapors.

Flammable range: Flammable range describes the percent of fuel in the gaseous state to air with 21% oxygen in which ignition can occur. Flammable range is expressed from the lower flammable limit (LFL), the minimum percent of fuel needed for combustion, to the upper flammable limit (UFL), the maximum percent of fuel that will burn in air. Mixtures below the LFL are too lean, meaning not enough fuel is present, whereas mixtures above UFL are too rich, meaning not enough air is present to support combustion.

Flash point: The rate at which liquids evolve vapors depends on the liquid's temperature. Increase of liquid temperature results in increased vapor evolution. The temperature where liquids produce ignitable vapors in sufficient quality to ignite is known as the flash point.

Autoignition temperature: The temperature where vapor/air mixtures within the flammable range will ignite without spark or open flame is the autoignition temperature of gases.

Minimum ignition energy (MIE): The amount of energy from a spark or flame needed to initiate combustion in ignitable mixtures of gas and air. This spark initiates combustion in a very small volume of the gases. Once ignition occurs, flame spreads through the remaining ignitable volume.

Boiling point: The temperature of liquid with maximum production of vapors at standard pressure. For water, this is 212°F.

Vapor density: Vapor density is a comparison of the mass/volume ratio of vapors to air and is expressed in a number. Any number lower than 1 indicates the vapor has less density than air, and thus will rise in air. Numbers greater than 1 indicate the vapor will sink in air. Vapor density is also known as the specific gravity of vapors.

Specific gravity: Comparison of mass/volume ratio of a liquid to water defines specific gravity (SG). Water is expressed with SG of 1. Liquids with SG less than 1 will rise and float on water, and those with SG greater than 1 will sink. *Note*: Some liquids are miscible in water, and therefore will mix.

Miscible: Liquids that mix with other liquids are known as miscible. Miscibility may best be understood with the old adage that oil and water do not mix. Petroleum hydrocarbons are not miscible. Alcohol, on the other hand, is miscible with water.

Flammable gas: A material that is in the gaseous state at standard temperature (20°C) and pressure (14.7 psi) and can form an ignitable mixture at 13% or less gas-to-air volume ratio or has a flammable range of 12% or greater.

Flammable liquid: Liquids with flash points at or lower than 100°F are known as flammable liquids. In most atmospheres these liquids produce sufficient vapors to reach the lower flammable range, and thus are extremely hazardous.

Combustible liquid: Liquids with flash points greater than 100°F are known as combustible liquids. Under normal conditions they do not produce sufficient vapors to reach LEL. *Caution*: That is normal conditions. When liquid temperature exceeds the flash point, such as liquid on asphalt pavement on a hot day, the liquids act exactly as flammable liquids. Additionally, sprays of combustible liquids (aerosols) ignite in temperatures lower than the designated flash point.

Ignition of Solid Fuels

Much like ignitable liquids, solids must change state of matter to vapor for flaming combustion to occur. The temperature at which solids produce sufficient vapors for those vapors to reach an ignitable mixture with air is expressed as the ignition temperature of that solid. No flammable limits are designated for solid combustibles.

Fire Extinguishers

Portable fire extinguishers provide "first aid" fire suppression, meaning they are intended for quick use until the fire is extinguished, the fire becomes dangerous to those using extinguishers, or the fire department arrives. Portable fire extinguishers are selected based on the hazard posed by fuels in the area to be protected. The classification system for fire extinguishers is extremely important to assure (1) the extinguishing agent works on the fire being fought and (2) more importantly, the fire extinguisher does not expose the user to danger on specific fires. An example of danger is using water to fight a fire involving charged electrical equipment.

The classification system is designated in letters and symbols, A, B, C, D, and K. These classifications are based on fuels and hazards involved (Figure 7.2).

Classification	Fuel or hazard
A	Ordinary combustibles including wood, paper, and plastics. These are characterized by char remaining after combustion.
B	Ignitable gases and liquids. Fuels that have no char remaining after combustion characterize these.
C	Electrical hazards–any fire that involved or is near energized electrical equipment is classified as C. Conductive extinguishing agents should not be used until electricity is disconnected.
D	Combustible metals. Examples include magnesium, aluminum, sodium and titanium. Care must be exercised to assure specific extinguishing agents are compatible with the burning metal. Generally the agent is rated for a specific metal not multiple metals.
K	Kitchen fires. Though oils used to cook food are ignitable liquids, they act differently after flames are suppressed than do other hydrocarbon fuels. Retained heat in these fuels results in reignition if proper agents are not used in suppression.

Figure 7.2 Fuels and hazards.

NFPA 10, *Standard for Portable Fire Extinguishers*, specifies travel distance to fire extinguishers and other criteria for selection and placement related to specific hazards. Maximum travel distance is a minimum requirement, and thus may provide insufficient access to extinguishers during emergencies. For example, for Class A fires the maximum travel distance is 75 feet. In a 100-foot-long and 50-foot-wide building, one centrally located extinguisher may meet the code requirement; however, the wisdom of requiring occupants to travel deeper into a building is questionable. Better distribution would require at least two extinguishers, one located near each of two remote exit doors. This allows occupants to choose between fighting the fire or egress. When choosing to fight the fire, they remain between the fire and their egress, which is much safer than having to move around a fire to reach an egress.

OSHA requires employees expected to use fire extinguishers must be trained in their selection and use. Having employees leave the endangered area and allowing a fire brigade to fight the fire or choosing against fighting fires is acceptable; however, fire extinguishers are required by fire codes regardless if they will be used or not.

Servicing and Inspection

NFPA 10 includes product-specific checks, servicing, and inspection requirements for portable fire extinguishers. Checks involve assuring extinguishers are properly placed, charged, not used or tampered with since the last service, and are ready for immediate use. Minimum training is required for those performing these checks that must occur monthly. Annual inspections require a bit more training to assure the unit has been serviced and tested within specified periods, in addition to all aspects of the monthly check. Testing, including pressure testing of the container, must be performed on a regular schedule specified by the type and construction of the extinguisher and must be completed by a qualified individual.

Fixed Fire Protection Systems

Most industrial facilities are equipped with fire protection equipment that does not require human interaction to initiate their operation.

These include fire alarm systems, fixed fire extinguishing systems, and fire sprinkler systems.

Fire Alarm Systems

Fire alarm systems use detectors to sense the presence of excessive heat, infrared heat, rapid rate of temperature rise, or smoke. When fire or fire products are detected, audible or visual alarm is made to notify occupants of a fire event. Most systems are connected to monitoring services that notify public fire departments to respond.

Fire drills, practice evacuations, are required in most occupancies by fire codes. Drills are intended to assure occupants recognize alarms meaning that an emergency exists and provide an opportunity to practice evacuation. Occupants must never act as though the alarm is false or involves a practice; rather, they must evacuate or go to proper areas of refuge in an orderly manner. The exception to evacuation is specially trained persons who are assigned to coordinate the evacuation or response activities.

Regular testing of fire alarm systems is required by NFPA 72. Inspections are necessary to assure all detectors and alarm components work properly. Qualified individuals who are trained on that manufacturer's equipment and NFPA 72 requirements must perform inspections, repairs, and alterations to the system.

Caution: Resist the temptation to silence alarms "until we know what we have." That practice has resulted in major life loss many times in history. If repeated alarms occur, a problem exists. Investigate the situation and repair the system.

Sprinklers

Fire protection sprinkler systems have demonstrated the ability to control fires quickly and prevent losses, including loss of property, but more importantly, they prevent loss of life. Historically, the primary reason for sprinkler failure is human intervention. Often, that is someone closing the valve before assuring the fire is controlled.

Sprinkler systems are designed under the specifications outlined in NFPA 13, *Standard for the Installation of Sprinkler Systems*. Design is based on a volume of water per area of floor space based on the fuel type and arrangement. Differences in design may mean a properly installed sprinkler system designed for a fuel load and arrangement may not be successful against another fuel or arrangement. Changes in process, equipment arrangement, materials stored, and arrangement of potential fuels must be evaluated by a licensed professional engineer or person certified to evaluate sprinkler systems. Many insurance companies offer services to assist in assuring proper protection is provided.

A common misconception of fire sprinklers is that when heat is detected, all sprinklers activate simultaneously. Though some systems work under this principle, called deluge systems, they are relatively rare and protect specific hazards. Most sprinkler systems work on the principle that when a sprinkler reaches a determined temperature, that head opens to flow water. If adjacent sprinklers reach the designated temperature, they too open and water flows. The number of sprinklers that may open before the system is intended to control fires varies, but rarely exceeds 20 sprinklers. Five or fewer sprinklers operated in approximately 90% of fires where they controlled the flames (Hall, 2013).

Resist the temptation to turn sprinklers off before the fire is located and a charged fire hose line staffed by trained firefighters is in place. Damage caused by fire suppression activities, including water damage, is usually covered by insurance. Insurance companies generally prefer to pay for water damage than to risk total loss that often occurs following premature sprinkler closure.

Anytime a sprinkler valve is closed during a fire event, a competent person with reliable communications must remain at the valve until relieved by the fire department incident commander. Preferably, the person closing valves is a firefighter assisted by a facility employee.

Annual inspection of sprinkler systems in accordance with NFPA 25, *Standard for the Inspection, Testing, and Maintenance of Water-Based Fire Protection Systems*, is critical to assuring proper sprinkler operation when needed. Schedules in NFPA 25 include time frames for more in-depth inspection of sprinklers, including assuring integrity of piping from corrosion (5 years) and assuring sprinkler heads operate properly (50 years).

Fire Extinguisher Systems

Specialized fire extinguishing systems are often installed to protect specific equipment or processes. Operations feature a device to detect specific conditions of temperature, rate of temperature rise, or infrared heat. Detectors cause activation of controls that release a fire extinguishing agent. Agents include dry chemical, wet chemical, carbon dioxide, halon, halon replacement, and other specialized agents.

Halon gas may still be found in specific situations, but is being phased out. Though it is legal to use agent that exists, manufacture of halon is not permitted by international laws. When used, the agent can only be replaced from stockpiles of nonused agent.

Each system has design criteria based on hazard posed and agent used. Changes in process equipment, rate, or materials used must be evaluated by competent individuals to assure systems will extinguish fires presented.

Semiannual inspection of fire extinguishing systems is required by NFPA codes. A person trained and competent with the equipment and arrangement must perform this inspection.

Hazard Evaluation Plans

Fire hazard analysis is needed for each facility and process. Competent fire protection professionals need to evaluate fuels that could potentially burn and heat sources to determine the possibility of fire occurring. Evaluation also includes potential fire spread to adjacent fuels and equipment. Determination also includes threat to life and potential injuries. Information is then evaluated and protective measures are designed to reduce potential threats.

Fire Protection Handbook

Published by the National Fire Protection Association, the *Fire Protection Handbook* is a two-volume reference to enhance understanding of fire prevention and life safety measures. It is updated periodically, with the 21st edition due in or around 2015. An electronic version is likely for this edition of the informative reference.

References

Hall, John. 2013. *U.S. Experience with Sprinklers*. Quincy, MA: National Fire Protection Association. http://www.nfpa.org/~/media/Files/Research/NFPA%20reports/Fire%20Protection%20Systems/ossprinklers.pdf.

8

Writing Effective Policies and Procedures

TRUETT A. RICKS, BOBBY E. RICKS, AND TRUETT GRAHAM RICKS

Contents

Organizational vision and mission give direction to its operations. This vision is critical to the success of a business. Policies and procedures create an internal system to ensure employees follow the objectives set to help follow the company's vision. Policies control the direction of employees and help establish goals for the business. For consistency, policies should be in writing. Written policies provide a consistent reference for employees to follow when striving to achieve objectives and reach goals. Unwritten instructions cause work to be performed inconsistently, thereby causing business interruptions.

The security unit must develop policies and procedures to promote the vision and mission while protecting organizational assets. A policy tells us what to do, while a procedure tells us how to do. A policy is the organization's guiding or governing principle. It is a general guideline that sets forth parameters for decision making along with the authority for implementation. Policy development is a planned process requiring due diligence in thought as a policy becomes a strong principle of the organization. Planners must look to the desired outcomes while considering the needs of end users. Policy justification should reflect problem solving while demonstrating persuasive reasoning, clarity, and coherence. How will the policy be enforced? Is there an incentive (including negative incentives)? Is there support for the policy from the workforce? Is there some overriding need for the policy? The policy should be consistent with the organization's mission, culture, strategy, and vision. It should not overlap or contradict with other policies and procedures.

Clarity is important in writing a policy. The average reader should not have any questions about the objective of the policies. When writing, do sufficient research to assure you are compliant with other policies and legal requirements. All stakeholders should have an opportunity to have input and give feedback about the policy. In a large organization, worker committees or teams may be organized to assure feedback is received from all levels. When written, policies must be published in a form that is easily accessible to those affected by the policy. In general, a policy manual should be easy to use with a format that allows for ease in updating.

A policy or procedure should be written (or revised) when there is confusion about the appropriate action to take in a given situation, for fairness, consistency, and accountability. Sometimes, legislative or regulatory changes require a change in policy or procedure. After a while, a policy or procedure may have so many exceptions, exemptions, or waivers that the policy (or procedure) is ineffective. This requires a review and either a rewrite or enforcement of the present policy (procedure).

Not all actions require or fall within a written policy. While policy sets out the order of business, anomalies occur and are handled on a case-by-case basis. Policy makes for efficient work. We may be tempted to write a policy for every possible event and become inefficient spending valuable resources writing nothing but policy, and you end up with a policy manual so large it won't be read. The manual becomes an organ to catch people doing wrong rather than direct people on doing right.

The policy should begin with a policy statement. The policy statement explains what you are doing. Is the policy a standard or a guideline?

Standards are specific requirements that must be met, while guidelines identify best practices. The statement tells when the policy applies and lists major conditions or restrictions. Next is the reason for the policy and a description of the conflict or problem the policy is designed to resolve. List those who should know this policy and follow the policy and accompanying procedures in order to do their job. Document the resource and contact information at the end of the policy.

Procedures tell us how to do something. They are the steps for getting things done. Inside of a procedure may be work instruction, which is specific steps within the procedure. Work instructions usually involve one person accomplishing the task. The aim for a procedure is to use one action per step and to assign the action and who is responsible for the action. Many of the general considerations of procedures occur when writing policies. In lieu of repeating material, just keep in mind those ideas presented above.

A procedure begins with a purpose statement. What are you trying to accomplish? Then list the actions and sequence, who does each step, to include handing off to another person, where and when this must be done, and any standards for completing the work.

When a new policy or procedure is adopted, the work population must be educated as to its existence. In the case of a procedure, this may require training. Whether through a training session or a briefing, notice should be formal, with documentation of all persons receiving the new material. New employees should be notified of policies and procedures in new employee orientations.

When the policy, or procedure, is in effect, enforcement is usually through disciplinary action. Discipline must be consistent or the policy (procedure) will lose its power. Depending on the severity of a policy violation, progressive discipline provides a level of punishment commensurate with the violation. For example, a first offense may result in a 1-day suspension; a second offense, a 1-week suspension; and a third violation, termination. In such a case, two employees could get a different punishment for the same conduct. Another consideration may be the seniority of employees. Discipline may be more severe for a seasoned veteran who should know the policy than a new employee. Some policy violations are so severe to warrant immediate termination. These are usually reserved for those violations that could result in death or serious injury to others, or on matters of integrity.

Embezzling money from the organization may not create a risk of injury, but the nature of the violation would warrant dismissal.

When a policy (procedure) has not been enforced, or has been selectively enforced, reviving the policy (procedure) follows basically the same notification process as a new or rewritten policy (procedure). When reviving a policy (procedure), personnel are formally trained with the addition that the policy (procedure) has not been enforced, that it is still applicable and necessary to the function of the organization, and that forthwith, the policy (procedure) will be enforced. Consideration should be given to a graduated disciplinary process in reimplementing the policy (procedure).

There are two types of security policies: policies affecting the workforce and policies specifically related to the guard force. Policies affecting the workforce are as varied as the imagination allows. Some of the most common are policies affecting access control, company equipment, visitors to the workplace, and information security. Policies affecting the general workforce are often seen as restrictive and a hassle. It is important for the policy to state the reason for the policy and that the policy be applied throughout the organization. If the CEO doesn't have to comply, employees begin to question why they have to comply. Executives should set the example by following all policies and directing others to do so.

Security Policies That Affect the General Workforce

Access Control

A policy on site access control (not IT access control) should state reasons for buildings and doors to be open or locked, and the time doors will be open or locked.

> The administration building doors will be unlocked Monday–Friday at 7 a.m. Access at other times is by key or contact security.

The policy should dictate who is entitled to a key and a log be maintained of all keys issued. The policy should require human resources to withhold final pay until keys are returned, for example:

> Keys are issued to all full-time employees for their office. Keys shall not be duplicated and are to remain in the possession of the employee at all times. Keys shall be returned upon resignation or termination.

Supervisors are authorized a master key for all offices in their area. Master keys shall not be loaned out and shall remain in the possession of the supervisor at all times.

Electronic access control systems should dictate times and days personnel are allowed to enter the premises. Access control cards and fobs shall be returned on resignation or termination. Employees should be aware that their card may be deactivated for security reasons, for example:

Employees are issued access control badges that shall be displayed on the person at all times while on the property. Badges should be visible above the waist by hanging on a lanyard around the neck or clipped to a shirt, blouse, or jacket. Employees are not to loan the badge to another person. Misuse can result in having your badge deactivated.

Visitors

Visitor control policies should state the process for permitting visitors access to the facility and who has permission to sponsor a visitor, for example:

Any employee may sponsor a visitor to the facility. Visitors must be met at the front desk by the employee who will sign for the visitor badge and escort the visitor at all times while he or she is in the facility. The employee is responsible for signing out the visitor when he or she leaves the facility.

Visitor policies may also direct areas where visitors are not allowed access, for example:

No visitors are allowed in the research lab at any time without the written approval of the lab director.

A visitor policy may contain a procedure employees will follow if they find an unidentified visitor, for example:

Any employee who observes an unescorted visitor or unknown person not displaying an employee badge shall immediately notify security and provide a description of the person. The employee is not to approach the unknown person. The employee may be asked to watch the person until security arrives and to notify security if the person moves to another location.

Company Equipment

A policy for equipment control should state the authorizations needed to remove property from the premises, for example:

> All company equipment is marked with an identification tag and must be checked out before leaving the facility. The employee must have the approval of a supervisor to remove property from the facility.

Information Security

Information security, whether hard copy or digital matter, should be secured by users and steps taken to guard against theft or compromise, for example:

> All documents shall be secured at the end of the workday by placing all files in the workroom. The workroom shall be closed and locked at the normal close of business.
>
> Documents shall not be left exposed on any desk or workspace. If you need to leave your work area briefly, use a coversheet to protect the documents from view.
>
> Computer screens should be positioned to protect against view from passersby. When leaving your workstation, computers should be placed on standby with a password to reopen.

Policies Affecting the Security Force

Jurisdiction and Authority

The security policy should include a policy identifying the jurisdiction and authority of the guard force, for example:

> XYZ, Inc. guards have no authority to arrest. State law [cite statute] authorizes guards on premise to detain suspicious persons for law enforcement authorities.
>
> Guards have no authority or jurisdiction to act off company property.

Code of Ethics

The security force should adopt or develop a code of ethics. All personnel shall receive training on the code of ethics and agree to abide by the code, for example:

> All security personnel will receive _ hours of ethics training. At the completion of the training, each employee will sign an agreement to abide by the code of ethics.

Use of Force

The decision to allow or not allow force by guards should be stated with details on the use of force allowed or authorized by law, for example:

> Guards are not allowed to use force except to prevent injury to another person. The amount of force permissible is dictated by state law [cite statute]. At no time is deadly force authorized to protect property.

When lethal or nonlethal weapons are issued or allowed, the policy should dictate the use of the weapon, for example:

> Firearms are issued for the personal protection of the guard and personnel on site. Deadly force is authorized by state law [cite statute] in the following instances:
>
> 1.
> 2.
> 3.
>
> Pepper spray is issued to guards for personal protection. Pepper spray is allowed when....
>
> If a guard discharges his or her pepper at an assailant, the guard must immediately inform the assailant he or she has been exposed to....

The use of force policy should require a report on any use of force, for example:

> When a guard must resort to force in the performance of his or her duties, the supervisor should be notified immediately. The guard will write a detailed account of the event, expressly including the following:
>
> 1.
> 2.

Use of Equipment

A directive should govern the use of equipment for official purposes only and only in the line of duty, for example:

> All property issued to the employee is for official use only and is intended to be used only in the line of duty when circumstances warrant its use.

Uniform Requirements

A directive should govern the uniform and what may or may not be displayed on the uniform, for example:

> The uniform is for duty use only. Employees are permitted to wear the uniform to, from, and while on duty. Only authorized badges and insignias may be displayed on the uniform.

Report Writing

A policy on reports should include when a report should be written and what should be included in the report, for example:

> Security reports must be completed when:
>
> 1. There is a violation of a security order or safety rule.
> 2. Someone reports a crime to security.
> 3. An alarm is activated.

Resources

Ricks, T.G. 2013. The Necessity and Evolution of Human Resource Related Policies in Small Business. Master's thesis, Midway College, Midway, KY.

PART II

Infrastructure Protection

9

Overall Physical Protection Program

JEFF DINGLE AND BOBBY E. RICKS

Contents

Generally, we can break security down into five levels: minimum, low, medium, high, and maximum. As levels increase, everything from the lower level is included in the higher level. For example, the policies, equipment, and procedures in the low and minimum levels are included in the medium level.

At the lowest level—minimum—are simple physical barriers designed to impede unauthorized external activity. The key word is *impede*. At this level, we are not trying to stop everything, but to discourage criminal activity. The second key term is *external activity*. At this level we are only going to worry about external activity—internal issues are not a concern at this level. The minimum level includes simple locks and simple physical barriers, such as doors.

The level low is simple physical barriers supplemented with additional devices intended to impede and detect unauthorized external activity, including using reinforced doors, window bars and grates, high-security locks, simple security lighting, and a basic local alarm system.

The medium level incorporates minimum and low-level measures, augmented with impediment and detection capability to impede, detect, and assess unauthorized external activity. The medium level includes the capability to assess unauthorized external activity. Note that even at the medium level, we are still only concerned with external activity. The medium level utilizes advanced intrusion alarm systems, established perimeter protection with penetration-resistant fences topped with barbed wire, or patrol dogs and unarmed watchmen.

The high level incorporates the measures of the previous three levels to impede, detect, and assess unauthorized external and internal activity. The high level includes formal contingency plans, local law enforcement coordination, high-security lighting, access control, highly trained armed guards, a perimeter alarm system, and a closed-circuit camera system.

Finally, the maximum level includes everything necessary to impede, detect, assess, and neutralize all unauthorized internal and external activity. This is the most secure level, and the most expensive to accomplish. This may include a sophisticated alarm system and an on-site, armed response force.

As security increases, more emphasis is placed on perimeter security.

Once again, as with any part of the physical security operation, it is important to identify what it is that you are trying to accomplish. Perimeter security is no different. Generally, perimeter security is the first line of defense; it is the first thing encountered by anyone approaching a facility.

From a security standpoint, for any facility, there are four methods to enter: forced entry, stealth entry, accidental entry, and entry by permission. A forced entry is exactly what it implies. It is a forcible entry, generally resulting in damage. It can be a hole cut into a fence, a window broken out, or a door kicked in. From a security standpoint, a forced entry is not necessarily bad. A forced entry shows proof of an entry, which lets us know we have a problem. A stealth entry, on the other hand, is a forced entry that leaves no evidence. This can be picking a lock or climbing a fence. A stealth entry is worrisome because there is little to no evidence left behind. No evidence means difficult to detect. An accidental entry is just as it implies. An accidental entry is made when someone accidentally enters an area. Typically, an accidental entry is not malicious but can cause other problems with people in areas that they don't belong.

Finally, a fourth method of entry is entry by permission. Entry by permission means that we have allowed someone to enter our facility. Remember, just because we have allowed access does not mean that the individual is authorized access.

Of the four methods of entry, controlling entry by permission is a function of access control, which is addressed elsewhere in this text.

Perimeter security techniques protect against forced entry, stealth entry, and accidental entry.

When creating a secure perimeter, a perimeter system will provide six distinct functions; for this text, all begin with the letter *D*: define, deter, delay, detect, direct, and decrease.

First, a perimeter will define the boundaries of the facility. This lets everyone know what is public property and what is a protected area. Definition can be important from a legal standpoint as well.

Next, a good perimeter system will deter a criminal from attempting to gain access. Ideally, a criminal will simply look at the perimeter system and decide it is too difficult to defeat or the likelihood of getting caught is too great. That is a proactive deterrent approach. A good perimeter system offers some level of delay. Many things can delay a person—having to climb a fence, cross a ditch, or get through a lock. Anything can be defeated; it is a question of how much time and effort someone is willing to take to defeat it. Remember, the more you slow down an intruder, the more likely you are to catch him or her.

For example, an intruder may attempt to enter your site through a storm drain. A manhole cover with brackets securing the cover from the topside does not make it impossible for an intruder to get through; it makes access more difficult and will slow down an intruder. From an operational standpoint, it also slows down maintenance personnel who need access.

An advanced perimeter security system will have detection capability. An intrusion detection system lets us know when protected space has been violated. It lets us know when a fence has been climbed or a gate has been opened. Intrusion systems on a perimeter will give us the greatest amount of time to respond, as the perimeter is the greatest distance away. Perimeter and intrusion detection will be discussed later in this text.

Lastly, perimeter systems allow us to direct traffic, both foot traffic and vehicular traffic, to a specific access point on the perimeter. Direction makes it easier for us to screen access, and makes it easier for the public to understand where they are supposed to go. This can be as simple as a temporary perimeter system made with 4 × 4s and snow fencing to both define and direct people to a checkpoint.

When we look at perimeter barrier construction, barriers fall into two categories: natural and man-made. A natural barrier is anything that naturally exists that accomplishes one of the six Ds. When utilizing natural barriers, first maximize the use of existing terrain. Small bodies of water can effectively be used to define boundaries and delay access. Even a simple ditch can be useful in keeping automobiles on roads.

Man-made barriers are any barriers constructed for use as a security device. This can include offenses, vehicle barriers, or walls. Sometimes a building's exterior may be its perimeter. Congested urban areas, such as a town center with skyscrapers and buildings set out to the sidewalks, and city planning and zoning regulations require you to get creative in protecting your facility. Some companies spend a great deal of money to create a high-security perimeter that is aesthetically pleasing and meets the needs of facility protection.

While standards vary, the generally accepted standard for a security fence is 11-gauge chain-link fencing material with 2-inch mesh. Gauge, the weight or thickness of the wire, is inversely proportional, meaning the smaller the number, the heavier the wire. For example, 9-gauge wire is heavier than 11-gauge wire. The fencing material should be a maximum of 2-inch mesh, meaning that the squares in the wire are no greater than 2 inches on each side. The fencing material should be a minimum of 7 feet high, with an additional foot of barbed topping, resulting in a total fence height of 8 feet. In sandy or loose soil, the fence mesh may be buried 12 to 18 inches below grade to prevent digging under the fence. Additionally, a clear zone of 10 feet on each side of the fence for a total of 20 feet is desired. The clear zone keeps everything away from the fence to prevent anything from assisting a person from climbing the fence, and to keep anything from providing something to hide behind.

While no fence is climb-proof, certain designs may make a fence climb resistant. Fences manufactured by welding wires closely can prevent finger or toe holds as a person attempts to climb. One manufacturer designed an arch to make it difficult for climbers to use both hands and feet to climb once they reach the arch. A downrigger extending from the top of the curve further discourages climbers.

Fencing can also be modified to deter vehicle intrusions. One method of preventing vehicle penetration is to add high-strength cable along a fence line. One cable should be placed 30 inches aboveground for car attacks, and the other cable 35 inches aboveground for truck attacks. Cables are securely fixed to a reinforced steel post and overlap where one cable ends and another cable begins. The cable is attached to the fence to maintain height, but the strength comes from the anchoring posts. Most fence posts are hollow and provide little support to deter vehicles.

Concrete highway barriers are also used as antivehicle barriers. Portable versions are made with high-impact plastic and can be rapidly filled with water to provide a quick barrier. There are many ways to increase the security of a perimeter system.

A better and more permanent solution is to install bollards to stop vehicle intrusions. Bollards can be placed on the outside of fences to protect the integrity of the fence in the event of a vehicle attack. The bollards can be decorated with linking chains between them to direct foot traffic. Bus stop seats and rain shelters can be made to act as a vehicle barrier. Constructed out of stronger material and properly anchored, these features add aesthetics to an area while providing protection. A more sophisticated vehicle barrier is a "tiger trap." A sidewalk is constructed over a loose foundation that supports foot traffic, but will collapse under the weight of a vehicle.

The building wall is often the first layer of defense. A concrete wall protects more than a metal-sided building. Interior walls of Sheetrock can be punched through with hand tools, affording little security once a perpetrator is inside your facility. Doors and windows should be visible with lighting so the burglar will not be concealed while he or she works. Most unlawful entries occur through the first floor, with the front door being the preferred method of entry. This does not reduce the need to secure windows and rear entrances—or the roof. Rear doors and windows are especially vulnerable as they are often out of public view. All exterior and some interior doors should be secured with locks designed to withstand forcible assaults or picking. Emergency regulations often require exit doors to open outward, exposing the hinges. Commercial doors are available with recessed hinges to thwart such attacks. Doors with exposed hinges should be pinned to prevent removal of the hinges and entry from opening the opposite side of the door. On interior doors, it does no good to have a strong, sturdy door if all the perpetrator has to do is break through a thin wall of Sheetrock. The same goes for interior locks. A high-security lock may look imposing, but it is useless to the attack on the wall. Interior areas with high-level assets should consider materials to delay entry: concrete or steel bars inside the wall that create an invisible "cage."

Windows can be coated with a safety glazing that reduces shattering. The glazing holds the glass together in an impact, slowing down entry. The glazing often has aesthetic properties like reducing glare and energy efficiency. Window locks should function. Test locks to determine the difficulty in manipulating them open. Weak locks should be replaced or reinforced. In some applications a window may serve as an emergency exit.

Vaults and Safes

Many types of storage equipment are used to store classified and sensitive information, weapons, controlled substances, valuable equipment, and negotiable documents or funds. The U.S. General Services Administration tests and approves security containers for the government, and these standards are useful in private industry as well. Containers are categorized as Class 1 through Class 6. The high end is a Class 1 security container insulated for fire protection. The protection provided is for 1 hour against fire damage to contents, 30 minutes against surreptitious entry, 10 minutes against forced entry, 20 hours against lock manipulation, and 20 hours against radiological attack. On the lower end of the scale is the Class 6 uninsulated container, with 20 hours against surreptitious entry, 20 hours against manipulation of the lock, and 20 hours against radiological attack.

Underwriters Laboratory (UL) tests safes before issuing a rating. The TL-30 rating is for a combination-locked steel container offering protection against an attack with common mechanical or electrical tools. The container must resist entry for 30 minutes. The TRTL-60 rating is for a combination-locked steel safe designed and tested to give protection against 60 minutes of attack and adds cutting torches to the list of tools. The TXTL-60 rating is for a combination-locked steel chest that offers protection against 60 minutes and adds high explosives to the list. A successful attack consists of opening the door or making a 2-inch square hole entirely through the door or body with a single tool or combination of methods.

A labeling service has been established by the Underwriters Laboratory to define the level of fire protection each safe can be expected to provide. There are three classes of fire-resistant safes. All three classes must pass three tests: fire endurance, explosion, and impact. During the fire endurance test, the inside temperature of a safe cannot exceed 350°F at any time during the test. At the end of the test, all papers inside a safe must be entirely legible and uncharred. A Class 350-4 safe withstands a testing furnace with the temperature raised to 2,000° at the end of 4 hours. A Class 350-2 safe must withstand 2 hours of exposure to heat reaching 1,850°F, and a Class 350-1 safe must withstand 1-hour exposure to heat reaching 1,700°F.

Information technology (IT) records begin to deteriorate at 150°F with humidity levels of more than 85%, so a container that has been described as a "safe within a safe" was designed. This container has a sealed inner insulated repository in which the IT material is stored and an outer safe protected by a heavy wall of insulation. This type of container has been designed to protect IT records against 150°F temperature and 85% humidity for the period specified. Insulated record containers are labeled by UL as Insulated Record Container, Class 150. An Insulated Record Container, Class 150-4 will protect the IT media for 4 hours.

10

Locks and Access Control

JEFF DINGLE

Contents

The building exterior often being the first line of defense, the most common entry points are doors and windows. Many security features help identify an intrusion, such as alarms and closed-circuit television (CCTV) systems. Without good locks, your alarm system will let you know when there is an intrusion and the CCTV will record the intruders taking your stuff. Locks and access control systems should allow authorized users while refusing unauthorized users. Proper systems should not let the unauthorized in or keep out authorized users.

Locks

Locks are useful in quickly securing (or opening) doors or access to things we want to protect. A lock is a mechanical device commonly used in a door, cabinet, or padlock. The commonly used pin tumbler, a lock with pins and keys to move the pins for opening the lock, was first patented in 1805. Different length pins are set in a tumbler, and a key moves the pins into position to allow the lock to open. They are not complicated—today's versions are basically the same.

Locks are mechanical, and as such, many locks can be defeated. The benefit of a lock is often not what it actually is, but what it is perceived to be. If a lock is "big and shiny," it may appear to be difficult to defeat. If someone has access to a key (even for a short time) and a duplicating machine, a key can be made that allows unauthorized access to your home, car, or business. Key control, secure storage, and documented issuing of keys are requirements of a good security program.

Locks fall into several categories, including key-in-knob, mortise, padlock, and electronic. Key-in-knob locks are just as described; the key cylinder is set into a knob or a lever handle. The key locks (or unlocks) an internal latch, and the knob is used to open the door.

A mortise lock is a locking mechanism, comprised of a large, rectangular body that fits into a mortise, or pocket, that is cut into the edge of a door. A separate keyed cylinder operates the locking and unlocking function of the mechanism. The mortise lock is generally used in commercial applications.

Padlocks are a small, self-contained lock with a shackle that is used to secure a hasp. Padlocks range from small "convenience" locks that are not very secure to heavy-duty high-security locks used by the government to protect classified material and weapons.

Electronic locks are locking devices opened or closed by means of electric current. Electronic locks have strikes electrically operated to open, or they can be magnetic with a strong magnet mounted on the doorframe that matches up with a corresponding armature on the door. Electronic locks can be stand alone, but are most commonly part of an access control system.

An Internet search will reveal several websites that provide an excellent look at a variety of locks and their strengths and weaknesses.*

Access Control

By definition, access control is the use of barriers and recognition devices to restrict access into a controlled area. Barriers are necessary, because if we are going to decide who we are going to let in, we must have a method of keeping out the people we don't want in. In order to utilize access control, the facility must first be secured. Fences, doors, locks, and other security devices are necessary to keep everyone out, so that we can use a central point to control access. The use of recognition devices, such as fingerprint readers, has become more prevalent in recent years as we try to identify an individual before granting access.

---

* www.blackhat.com/presentations/bh-europe-08/Deviant_Ollam/Whitepaper/bh-eu-08-deviant_ollam-WP.pdf.

There are only four methods of entry into a facility: force, stealth, accident, and permission. Force, stealth, and accident are controlled through the use of a perimeter security system, while entry by permission is a function of access control.

Access control systems incorporate two separate but equally important parts to be effective: (1) equipment and (2) policies and procedures. Many companies have excellent access control policies and procedures that are often not followed. Good policies are totally useless if they are not followed. Likewise, equipment, such as a swipe card reader, is useless if not installed and used correctly.

In order for access control to take place, there are two fundamental requirements: a secure area and an access list. First, the area being controlled must be completely secured. It is useless to place an access control system on the front door when the back door is left unlocked. Second, an access list must be provided listing who is authorized access and any restrictions on access, e.g., not permitted on weekends. Who gets access to what is a *business* decision, not a security decision. Security does not operate as the access decision maker. From a security standpoint, we allow access to those on the access list. It is the function of the security department to enforce the access list, not to decide who gets access.

When defining access control operating systems, there are four fundamental systems to control access. Each of the four types of operating systems has its pros and cons.

The first way to control access is through personal recognition. Personal recognition means that access is controlled by an individual who must personally recognize you in order to allow access. The advantage of personal recognition is that it is not machine based, and therefore not affected by power failures or technical problems. The disadvantage to personal recognition is that it can only be successful when used with very small, very stable groups—groups in which the members do not change frequently. This type of system is more common than you might think. Outside of the office of the CEO at every large company sits a receptionist/secretary who has a great deal of control over who enters the CEO's office. While not officially an access control system, this is a common and effective personal recognition system.

The second system is unique knowledge. Unique knowledge means that in order to gain access, you must know something such as a code, sequence, or password. Unique knowledge might be the combination on a safe; in order to open the safe, you must have the unique knowledge of the combination. There are several problems with unique knowledge: First, if you share it, you cannot take it back. If you tell someone your password, you cannot take back that information. Second, if you give out the information, you cannot control who else the information is given to. If I have your password, you have no control over who I give it to. Passwords are often easy to guess, and many people use the same passwords for multiple applications.

IT applications recommend increasingly difficult passwords so that they are more secure, harder to "break." Passwords should use upper- and lowercase letters, numbers, and special characters. IT applications also recommend that for security reasons, passwords should be changed frequently. The combination of multiple characters and frequent changing make passwords difficult to remember, and as a result, users may write them down. A password that is written on a desk blotter, on a yellow sticky note, or commonly, under the keyboard is a unique knowledge that is not very secure.

The third operating system is unique possession. Unique possession requires that you possess something in order to access something. The most common form of unique possession is a key, but it may be a token or other device. These devices, however, do not identify the user. Any person with a key to a door can open the door. A person with the key to a car can start the car. The car does not know if you are authorized to have a key, or if you have copied or stolen the key. An access card or "swipe" card is also an example of unique possession. Access cards are often synonymous with identification badges. Again, an access card does not identify the user. The access card identifies who the card is issued to, not who is using the card.

The fourth system is unique biometric. Before allowing access, we identify the user by comparing existing biometric information to biometric information that is stored on file. A biometric is a unique, body-based identifier that cannot be copied or shared. Biometric systems *first* identify the user and record the information in the system. When the user attempts access, the system compares the input against the access list, and allows access if the information matches. Biometric systems include devices that read fingerprints, different parts of your eye, such as your retina and iris, the blood vessel pattern in your finger, or many other technologies. Facial recognition, or imaging and "mapping" your face, is an emerging technology.

Hand geometry is a common system as well. This system measures the width of the hand, the thickness of the hand, the length of the fingers, the thickness of the fingers, and the overall size and shape of the hand. Ten years ago hand geometry systems were a very advanced technology. Today, hand geometry readers are commonly used in business applications. This system is so common now that it is currently in use at many universities as an access system for student dormitories.

The combined use of multiple systems increases security by requiring the user to use two or more methods. A hand geometry reader may include a keypad where the user will enter a pin number (unique knowledge) while scanning the hand (unique biometric). Multiple concept systems exponentially increase the security—while defeat or bypass of a single system may be possible, defeating multiple systems simultaneously is less likely.

When utilizing an access control system, only four responses are possible. Either you are on the list or you are not, and either you are allowed access or you are denied access. If you are on the list and allowed access, that is good; that is what is supposed to happen. If you are not on the list and not allowed access, that is good; that is what is supposed to happen. However, if you are on the list and not allowed access, that is bad, but that is something that can be fixed. If you are not on the list, and you are allowed access, that is very bad. From a security standpoint, we would rather keep someone out who belongs than allow someone in who does not belong.

	ACCESS ALLOWED	ACCESS DENIED
ON LIST	GOOD!	BAD
NOT ON LIST	VERY BAD	GOOD!

System Integration

As system integration progresses, outside systems will have an impact on your access control system. For example, access doors with electronic locks may unlock during a fire alarm. It is important to know how your access system integrates with other systems in your building, and develop a plan in the event locks are released.

Access Systems

Common access control systems include key-and-lock systems, biometric systems, card reader systems, proximity systems, bar code systems, and keypads.

When selecting an access control system, several factors have to be considered.

Throughput time is the amount of time required to successfully use the access system. How much time does it take to approach the system, use the system, and ultimately gain access? Timing for recognition and access takes only seconds.

Error rate is the rate at which the system makes mistakes. Again, we would rather keep someone out who belongs out than let someone in who doesn't belong. Any system that has been in use for a length of time can compute an error rate. Imposter resistance is the difficulty in which the system can be fooled. A key has an imposter resistance of zero (because anyone with the key can use it), while a thumbprint reader has a very high level of imposter resistance, because it requires the designated person to be present. User compatibility and user acceptance have to do with the ease with which a system can be utilized, and whether people are afraid of the technology. Other considerations include the amount of time required to enroll new users into a system, and the amount of secured data storage required.

Cost is always a factor, but available money is determined by threat and need, and in the end, access systems must be reliable.

Locking systems and access control systems are critical to a successful security program. Keeping unwanted people out, while at the same time allowing easy access to authorized persons, makes for a professional operation.

11

Intrusion Detection Systems

BOBBY E. RICKS

Contents

Intrusion detection systems (IDSs) are designed to detect the entry of unauthorized persons into a protected area. The IDS should alert the response force when there is a breach of the perimeter. Building walls and the perimeter protect and delay the entry of attackers for a time. Openings—doors, windows, vents, and some walls—are the weak points in a security system. Intruders, even highly skilled intruders, look for the easiest way into a facility. A facilities security system should protect the assets until response forces can arrive. The earlier the notice, the more time for the response force to arrive and contain the situation. An effective IDS will complement your overall security plan. Countermeasures should reduce risk to an acceptable level at an acceptable cost. From one perspective, the earlier the warning, the less you have to depend on the perimeter and access control systems to protect your assets, but your thinking is generally focused on how to secure the most for less. The basic components of an IDS are a sensor, an annunciator or alarm, and a transmission method from the sensor to the alarm. The response force completes the system. An alarm does no good if no one responds. This chapter will explore commonly available systems and discuss the uses of each system. Response forces are covered in Chapter 14.

When there is an alarm, there are three possibilities: a valid alarm, meaning an intruder is detected; a false alarm, an unknown alarm that is not attributable to an intrusion; and a nuisance alarm, an alarm that is known and attributable to something like wind moving a sensor or vehicle vibrations setting off a sensor. Systems are: local, a system where the alarm terminates on site, usually with a bell or audible alarm to scare off intruders; proprietary, where the alarm terminates at an on-site or in-house monitoring station, such as a permanent security post that is also used to monitor closed-circuit television (CCTV), traffic, and cover telephone calls; and a central station system that terminates at an alarm company site, and the alarm company notifies response forces. Intruder systems are often coupled with fire alarm systems and alarm in the same fashion: local, proprietary, or central station.

Sensors function in different ways: a sensor may operate by connecting or breaking an electrical circuit, such as a magnetic contact at a doorway or window. When closed, a magnet holds an electrical contact in place. When the door or window is opened, the contact closes to connect an electrical circuit, setting off an alarm, or opens an electrical circuit, setting off an alarm. A photoelectric sensor uses a light beam to alarm. An opposed photoelectric system detects objects passing through a light beam, commonly used with automatic garage door openers. If the beam is blocked, the door will not lower. In an IDS, when the beam is blocked, an alarm is activated. Some photoelectric systems operate in the opposite manner, activating when the light beam transmitted reaches the receiver. Sound detection systems use a microphone to detect sound even the human ear can't hear. The system can be adjusted to ignore common sounds, thus reducing nuisance alarms. Vibration systems measure motion through changes in energy by velocity, acceleration, and displacement of that energy. The motion sensors are designed and calibrated to detect vibrations and alert when a pattern is disturbed or there is a pattern out of the ordinary. For example, a system may be programmed to ignore vibrations from constant vehicle traffic, but to alert to intermittent but steady vibrations, such as a person using a shovel or hammering on a wall. Other systems alarm when there is a capacitance change in an electrostatic field, such as microwave. Microwave transmission intensity changes when the field is interrupted or reflected away from the receiver.

Planning for the IDS starts with your security survey. Site and building diagrams help you design your security zones and determine the type of system needed to protect your assets. Zones of protection allow for one area to be used while other areas are alarmed. The same for building interiors—one floor or room may allow access while other areas are protected by the IDS. Identify the areas with the greatest ease of entry: doors and windows. Ground-level floors are more susceptible to entry because of ease. Is there roof access? Is there limited access to the grounds so an exterior IDS can be used?

Interior Systems

Doors are where intruders start. Doors can be alarmed with a mechanical switch that operates on an electrical circuit. The switch opens or closes to trigger an alarm, usually a button-type switch that depresses the button when the door is closed. When the door is open, the switch releases and triggers the alarm. In a closed-circuit system, the circuit is closed while the door is closed. When the door opens, the circuit opens, breaking the flow of electricity and triggering an alarm. An open-circuit system is the opposite. The closed door opens the circuit. Opening the door closes the circuit and triggers an alarm. Closed-circuit systems protect against intruders cutting wires in an attempt to deactivate the switch. Cutting the line opens the circuit and sets off the alarm as if the door had been opened. An open-circuit system would not recognize the cutting of the wires.

Magnetic switches operate in a similar manner: using magnets to depress and release the switch. The switch is in two parts: a magnet and a spring-driven switch. One part is secured to the door and the other part secured to the doorframe. The two parts are aligned so the spring-driven switch is moved when the magnet is in place. The alarm triggers when the door is opened, moving the two parts away from each other. Balanced magnetic switches use the same concept, but the magnets must be aligned precisely to activate the switch. Any imbalance will trigger an alarm. Both mechanical and magnetic switches provide trouble-free service with few nuisance alarms. They do require hardwiring, which can be costly if you have many entry points. These systems also work with windows, with the same considerations as a door. Also, the system would not detect someone already in the building, such as a person who hides in a facility until it is closed and secured. This makes it a feasible alternative when you need to secure a facility while personnel are present.

Motion detectors would detect a person hidden in the building—when he or she moves. This is a highly effective protection system against concealed intruders. Motion detectors emit energy into a room and alarm when the energy pattern is disturbed. Sensors use ultrasonic or microwave energy to detect the Doppler shift in the transmitted and received frequencies when motion occurs in the protected area. Ultrasonic systems transmit a pattern of acoustical energy, with a receiver picking up the sound pattern and establishing a base pattern. An alarm occurs when motion in the area sends back a reflected wave with a different frequency from the base pattern. Nuisance alarms can be triggered by loud noises, ringing telephones, machines, and other equipment that give off sound. Microwave systems operate in the same manner: they transmit a pattern of radio waves that are reflected back to an antenna. Stationary objects return reflected waves at the same frequency. Moving objects will return a different frequency, triggering an alarm. Microwave may penetrate walls and windows, triggering nuisance alarms by running water in plumbing or exterior movement through a window.

Passive volumetric motion sensors use infrared thermal energy patterns in a grid pattern, commonly known as passive infrared (PIR). The sensors alarm to changes of temperature moving through the field. The pyroelectric sensors are calibrated to make the system sensitive to the temperature of the human body. The system recognizes normal fluctuations in temperature, which are more gradual than a person moving through an area. The rapid change in temperature will trigger an alarm. PIR does not penetrate glass, but can be defeated by extremely slow movement. PIR sensors will have a light to assist in setting up the coverage field. This light should be turned off after setting the sensor so potential intruders cannot mark the limits of the field and avoid the area. PIRs should overlap to avoid blind spots in their coverage.

Photosensor motion detectors use focused light directed to a sensor at a specific frequency. A specific frequency keeps someone from using another light source to mimic the beam. Filters make the light invisible. When an object crosses the beam, light contact with the sensor is lost, triggering an alarm. The system is useful in open portals and driveways, and can be integrated to activate CCTV cameras. The effective range of photosensors is approximately 500 feet indoors and 1,000 feet outdoors. Mirrors can be used to bounce light in a grid configuration to increase coverage. There is a 10 to 30% loss of effectiveness for each mirror used in an array. When installing a photoelectric sensor, consider the beam path to obstructions such as tall grass, snow, and weather elements, such as fog, dust, or rain. Mirrors or multiple light sensor combinations keep an intruder from crawling under or stepping over a known beam path.

Sound detection uses microphones installed on walls, ceilings, and floors of the protected area. Sound detection is effective in enclosed areas, vaults, warehouses, and similar areas. The sensitivity of the microphones can be adjusted to meet operational needs. While this system is economical and easily installed, it is only useful in areas with a minimum of extraneous sounds. Glass-break sensors are tuned to the frequency of the shearing of glass. The frequency of breaking glass is distinctly different from other sounds. Vibration detectors are very similar to sound detection, effective in enclosed areas, vaults, warehouses, and similar areas. The sensitivity of the sensors can be adjusted to meet operational needs. While this system is economical and easily installed, it is only useful in areas where a minimum of vibration exists.

A final system is a capacitance or electrostatic system. An electrostatic field is set up around the object to be protected. The system is flexible and easy to set up, and creates an invisible field that is hard to detect. The capacitance of the human body will cause an imbalance in the electrostatic field, setting off an alarm.

Exterior Systems

An exterior IDS detects intruders before they reach the facility, increasing the time for the response force to arrive. There are several types of fence disturbance sensors. Shake detectors sense movement of the fence. Some sensors use free-floating contacts that move and touch when an intruder moves the fence by climbing, cutting, or lifting the fence. Shake detectors must be securely attached to the fence to reduce nuisance alarms. The detectors are vulnerable to animals touching the fence and to wind nuisance. They do not detect if a person tunnels under, bridges over, or jumps the fence. Taut wire sensors alarm when an intruder disturbs the tension of the wires by climbing over or moving the fence or by cutting the fence. They do not detect tunneling, bridging, or jumping and are affected by wind and animals. Some applications place taut wire inside a fence to reduce nuisance alarms by animals touching the fence.

Invisible barrier sensors commonly used in exterior applications are photo-light sensors, microwaves, ported or "leaky" cables, fiber optic cables, and seismic sensors. Photo-light sensors were explained above under "Interior Systems"—the facts are the same for exterior applications. Seismic sensors use motion sensors to detect ground vibrations. Sensitivity can be set to distinguish normal vibration patterns, with the system alarming when the pattern is exceeded; for example, the system may recognize foot traffic but alert when vehicle traffic approaches. Some sensors use electromagnetic fields that are interrupted by metal objects and produce an alarm. These systems are inexpensive to install and maintain. They are used in traffic applications to detect vehicles at traffic intersections, to change the traffic signals when they identify vehicles present at low-use intersections.

Fiber optic systems operate similarly to a taut wire system. Signals sent across the cable are measured for "strain." Changes in strain occur when there is a disturbance. The system alarms when a level of strain varies from the normal. The systems have little nuisance from wind, lightning, and other electrical disturbances. They can be deployed under gravel or sod, making application across an open lawn possible for early warning of potential intruders.

Ported or leaky cables are cable antennas that transmit a signal to a receiver. They alarm when a set field is disturbed. These cables can be buried to detect persons walking through the field. Depending on the space between the transmitting and the receiving cable, the field can radiate high enough and wide enough to prevent jumping over the field. Since the cables are buried, intruders can't see the field and determine tunneling or height/width of the field for compromise. Both fiber optic and ported coax are adaptable to uneven terrain. Ported coax can be affected by lightning and running water.

Microwave systems use a transmitter and a receiver to establish a detection field. The systems use bistatic multipath sensors to achieve a three-dimensional volumetric detection field of length, width, and height. The primary signal is sighted directly from the transmitter to the receiver. A secondary signal is a reflected signal bounced between the transmitter and the receiver. Bouncing multiple signals at different angles creates a low-level field that extends upward in a parabolic curve, the apex of the parabola being halfway between the transmitter and receiver. This configuration leaves a blind spot immediately in front of both the transmitter and receiver. This spot can be eliminated when overlapping transmitters and receivers in distance applications. A 30-foot overlap is recommended at intermediate points, and a 15-foot overlap at corners. Fences, walls, and other obstructions can create multipath signals, distorting the signal and possibly losing coverage on the back side of the obstruction. The greater the length of the microwave field, the wider the field becomes. Distortion caused by the width can be controlled by shortening the distance between multiple transmitter/receiver units. For example, a 600-foot pattern is 40 feet wide, while a 350-foot pattern is 20 feet.

The microwave system alarms when the pattern is disturbed, causing the energy to be reduced at the receiver. The system detects large objects, so the detection field must be located in an area free of obstructions. Many users apply microwave inside fencing to reduce nuisance alarms from accidental entry to the field by animals. The clear area should be as large as the protection pattern. The system may be defeated by tunneling or trenching. The parabolic field height discourages attempts to bridge over, but keep in mind that the curve is lowest closer to the transmitter and receiver. Overlapping patterns reduce this gap. Microwave systems can be stacked to raise the parabolic field to a greater height to eliminate bridging or jumping over the field. The initial set up is 2½ to 3 feet above the ground. Stacked systems place the second level at 5 to 6 feet aboveground.

Duress/Robbery Alarms

Duress alarms are activated by the user when a specific event occurs, or doesn't occur. For example, a guard may be required to enter a code or swipe an ID card hourly to prevent an alert from being activated. Usually, the alert is after an event such as a robbery or unwanted visitor. A "panic" button is located under a desk for an employee to request assistance. In some applications a floor-mounted kick pad is used, such as at a teller window in a bank. The teller standing behind the counter only needs to move the toe of his or her foot to activate the alarm. The button or foot pedal is usually encased in a housing so users don't accidentally set off the alarm.

Bait pulls are placed in a cash drawer and activate when all of the bills are removed from one bill slot. Removal of all bills allows contact to be made, thus setting off the alarm. In some instances, a "bait pack" will be placed with the bills to assist in tracking the robber. Some packs are activated once the robber leaves the area, while others activate immediately. Older packs emit a dye unique to security applications, making it useful in identifying the spoils taken in the robbery. New packs use GPS technology to track the whereabouts of the money.

Door and safe combinations and security codes can be programmed with a panic feature. Similar to password technology that locks a user out of a computer after a specified number of tries, a panic code is usually the user password with a key number in the code. For example, all codes in a system use every letter but *s*. The user under duress would only need to type an *s* in the code to create an alarm.

12

Security Lighting

BOBBY E. RICKS AND TRUETT A. RICKS

Contents

Lighting does two main things: enhance observation and deny camouflage. It is used to illuminate interior and exterior areas and strategic locations such as entrances and loading docks. Lighting applications consider safety and security needs in determining what to light and how much to light. There are different types of light sources available for security applications. When choosing the type of light, considerations include the cost, coverage, and service. You should also consider whether the lights are going to be used with closed-circuit television (CCTV), and whether the CCTV is black and white or color. Contrast and color rendition usage will be affected by the light source selected. The power source should be configured to have lights on separate circuits. That way, if a circuit fails, you will not lose all lights. In high-risk areas, lights should be connected to the emergency generator.

Light is measured in foot-candles, lumens, or lux. One foot-candle is the amount of light needed to illuminate 1 square foot from 1 foot away. One lumen is the same as 1 foot-candle. A lux is the measure of light striking a 1 square meter surface on which light is evenly distributed. One foot-candle or lumen roughly equals 10 lux. Light sources give off light in varying amounts according to the strength of the light and the absorption of the materials surrounding the light. A rough, black surface such as pavement will absorb light more than white concrete.

Types of Lighting

Security and facility managers work together to achieve sufficient lighting at an efficient cost.

Incandescent lights are inexpensive and easy to install, but they have high energy costs compared to other types of lights.

Fluorescent lights are more efficient, with efficiency increasing with longer tubes. Fluorescent lights do not function well in cold temperatures, making them most useful indoors. Metal halide lights are most color accurate, but have a long restrike time. Restrike time is the time it takes to restart a light when turned off. The safety/security will be compromised until the light restarts and resumes full capacity. If used for safety or security purposes, metal halide should be coupled with another light source with instant restrike time. While the initial cost is high, they have a long life and fair energy efficiency, which offsets the high initial costs. Metal halide is one of the best light sources when you need light for color CCTV. Low-pressure sodium lights are very energy efficient and have instant restrike. They have poor color rendition but good contrast, which may be acceptable in black-and-white CCTV applications. High-pressure sodium lights have excellent life and immediate restrike capability with poor to fair color discrimination and high life cycle cost. Mercury vapor lights are the oldest of the high-intensity discharge (HID) light sources. While you may have mercury vapor lights on the property, it may be cost-effective to upgrade to a more efficient lighting source.

In considering the cost of lighting, you must look at the initial cost and the cost over the life of the unit. Many HID lights have a high initial cost, but lower life costs than units with a low initial cost. For example, a mercury vapor lamp costs $26 and uses $720 in electricity over a 5-year period, while a low-pressure sodium lamp emitting the same light will cost $72, but will only use $187 in electricity for the same 5-year period.

Use the following formula to determine the actual cost of a unit:

Number of watt output × hours = watt hours	60 × 500 = 30,000
Watt hours ÷ 1,000 = kilowatts	30,000 ÷ 1,000 = 30
Kilowatts × rate = cost	30 × 0.06 = $1.80
Cost/life = cost per hour	$1.80 ÷ 500 = 0.003

With that said, the latest technology for security applications is light-emitting diode (LED). LED light sources are amber, red, green, and blue, but can be mixed or covered with a phosphor material to adjust the color of the light to white, which can produce color rendition close to normal daylight. A 13-watt LED light will emit 650 lumens, the same as a 40-watt incandescent bulb. The incandescent bulb has a life span of 750 hours; the LED will operate for more than 50,000 hours with a 30% loss of efficiency. The LED will produce 85% less carbon dioxide. All light sources deteriorate with time and use. A heat sink pulls heat from the LED, which extends the life and strength of the light. A retail store chain found that replacing outdoor security lighting with LED technology recovered the cost of use and replacement in less than 2 years.

For security applications, LED lights provide excellent light per watt and color rendition. LED lights can be used for infrared illumination in night vision.

Light Survey

The security manager should survey the property grounds and building interiors to determine security and safety needs. Parking lots and sidewalks should be lit with no dark spots. Exterior lighting in parking areas and walkways should have a minimum of lighting for pedestrians to recognize hazards while walking and to be seen by vehicle traffic. For security applications, you should be able to recognize a person's face from a distance of 25 feet. Open areas, perimeter fencing, and building exteriors should have a minimum of 5 lux; building entrances, a minimum of 100 lux; and gatehouses, a minimum of 300 lux.

Trees, bushes, and other obstructions may diffuse light and produce dark spots. Parking areas for trucks and trailers may require taller fixtures to assure full coverage to eliminate dark spots caused by the vehicles. For new construction, lighting engineers lay out a design of fixtures, spacing them to assure a minimum coverage over an area. The number of light fixtures and spacing depend on the light source installed. For existing applications, using a light meter will assure a consistent, minimum coverage over the area to be protected.

Light meter readings will be higher under the light source. Begin with the meter under a light source and move away from the fixture.

As you move away, the reading will drop. The lowest reading is usually at the midpoint between fixtures. The lowest light reading should meet or exceed the minimum requirements for safety and security needs. For example, if the minimum requirement is 5 lux, the lowest reading in the covered area should not go below that. Take measurements in a grid pattern over the covered area to assure consistent coverage. Note areas where coverage falls below the minimum standards. Additional lighting may be required to correct deficiencies. Rather than the expense of replacing an old and less efficient light source, it may be wise to consider a partial upgrade to correct the deficiency. Locations with new light technologies have discovered a greater area of coverage with a new unit. In new applications, fixtures can be spaced accordingly.

Make sure your lights support closed-circuit camera needs. Color cameras need a light with good color rendition, like metal halide or LED. A black-and-white camera will benefit from the high contrast of low-pressure sodium lights. LED technology can be specified for use at night with infrared cameras. Your lighting requirements need to specify what is needed for effective camera usage. Consider also lights for dark areas even in daylight or indoor applications.

Interior applications have the same considerations as outdoors. Hallways and work areas are lit with a minimum of light for safety and security. Emergency lights and backup electricity should be considered for all security applications and should be a necessity in high-security areas. Power to the lights should be on multiple circuits so if one circuit blows, the others will remain lit. Alternating lights on the circuits will prevent a large section of light failure. It is better to have some light in all areas than no light in one area. Outdoor lighting should be fitted with photoelectric cells to turn off lights during daylight hours. This reduces cost and increases the life span of the lights.

Not everything has to be illuminated. Dark areas require trespassers to bring their own light sources. One school that was plagued with graffiti artists realized that it was providing the light for the vandals. By turning off the lights, police and security forces could identify when the vandals were at work by seeing lights where there were not supposed to be any lights. The vandalism rate dropped.

Another "accidental" but effective use of lighting for security purposes is low-pressure sodium lights. The lights give off a yellowish color that leaves everything monochromatic—the lights make everything black and gray. Roadside rest areas have found reduced incidents of assaults because the light makes people look unattractive. Areas plagued with car cruising have placed low-pressure sodium lights on streets to discourage cruising. The cars don't look shiny and your date looks dead. In contrast, metal halide is used at auto dealers to show off the shine on new and used cars.

Light Ordinances

Many communities are enacting light ordinances designed to reduce "light trespass." Light trespass is when someone else's light is affecting you or light in excess of need. Bright security lights that shine on a neighbors' property can be annoying. Security lighting must comply with these ordinances. An example of a light ordinance is in the table below:

AREA	BASIC	ENHANCED SECURITY
Parking lot	4 foot-candles	7.5 foot-candles
Building	0.5 foot-candles	2 foot-candles
Entrances	10 foot-candles	10 foot-candles
Open areas	0.5 foot-candles	2 foot-candles
Pedestrian walk	4 foot-candles	6 foot-candles

New technology designs have full cutoff (FCO) light units that direct light in specific areas and reduce ambient light to meet light trespass ordinances while providing adequate coverage. The focus of the light increases efficiency, which lowers cost through fewer lights or by reducing power to the lights while achieving an acceptable level of lighting.

For Further Information

The Illuminating Engineers Society (IES) has published guidance on security lighting that is general in nature. Other sources of guidance include the Department of the Army Field Manual 19-30 and current Nuclear Regulatory Commission (NRC) regulations for minimum lighting levels.

13

Closed-Circuit Television Systems

BOBBY E. RICKS

Closed-circuit television (CCTV) is the premiere security tool. Due to the affordability, flexibility, and overall availability, video surveillance has evolved into an effective supplement to providing improved security. Since the creation of this visual tool in the mid-1900s, the overall theories and applications for CCTV remain the same. It is the tool that never sleeps, calls in sick, or takes vacation. With CCTV being all the rage, it is not the panacea for security operations; it is a tool that must be integrated with other security hardware and personnel to provide the desired level of security. Video surveillance reduces manpower by covering more areas with fewer guards, and emerging technology can alert the monitoring guard to an event in progress and allows security to assess situations as they develop. Recording the video provides a record of the event to be used in legal or business-related proceedings.

A major concern over CCTV systems is invasion of privacy. U.S. Supreme Court cases define privacy concerns where a person reasonably expects privacy and where society is willing to grant that expectation. There is no expectation that a person in a public place is not being observed physically or through CCTV monitoring. Banks, retail stores, government facilities, parking lots, hospitals, and industry use CCTV for security and other functions. Monitoring a person in an area where there is no expectation of privacy is not an invasion of privacy. Monitoring a person in a private area is usually considered an invasion of privacy. Signage at stores and private facilities warn people they may be under surveillance, thus lessening their expectation of privacy even in private areas. Employee manuals and written notice to visitors also put people on notice of CCTV usage, lessening their expectations of privacy. A person has no expectation of privacy in a place where he or she has no right to be. A burglar caught on video in a residence cannot claim a violation of privacy because he or she was captured on video while committing the crime.

Federal and state laws prohibit the interception of voice communication. Most CCTV applications are video only, so there is no problem with violations of eavesdropping statutes. State law and local laws may dictate areas or conduct that may not be recorded. Laws may prohibit placing a camera in a dressing area where a person may be intercepted in a partial state of dress, such as bathrooms, tanning booths, locker rooms, etc. There may be restrictions on placement of CCTV that may intrude on another person's area of privacy, such as a home system where a camera could capture images from other residences.

The basic elements of a CCTV system are a camera, monitor, and transmission method. The system can be integrated with other technologies, such as access control, intrusion detection, and recording systems to complement and extend the capability of those systems for improved security. The security manager doesn't need to be an expert in the technology, but should have an understanding of these components and develop a plan to incorporate the CCTV into the overall security plan. CCTV placement begins with taking the survey results and deciding how to apply the system: application, surveillance location, lighting, camera location, camera type, camera installation, monitoring system, monitor location, transmission method, and integration with other security applications (access control, intrusion detection, and recording). As you decide how and where to place cameras, additional considerations include the camera environment (exposure to elements, lighting conditions), mounting (pole, wall), system use (active or passive monitoring), remote access, and monitoring locations. The key question is to determine what you are trying to achieve—does the application of CCTV improve your security objectives? When discussing CCTV applications, describe your objective in what you are trying to achieve rather than technical jargon. For example, describe the application as: to identify persons who have access to a limited control area from a remote location, the camera should be able to see the complete entryway. If you told the installer you want a 6 mm lens with a 25° field of view, and the application did not achieve your objective, modification of the system to correct the error is an additional cost that could be avoided.

Decide what you are trying to capture. Coverage can be for detection, recognition, or identification. Detection allows you to see a man-sized mass on a monitor, and possibly movement. Detection is used in applications where you need up to 150 feet of coverage. Recognition can discern human attributes such as relative height, clothing color, and possibly see direction of movement. Recognition is used in applications of less than 30 to 75 feet. Identification can see unique human characteristics sufficient to establish identity and possibly hand movement. Identification is used in applications of 30 feet or less. Camera direction and light must be considered. On exterior applications, position the camera to avoid direct light into it. The camera will pick up the bright light and present your subject as dark and often undistinguishable. Changes in sun position throughout the day and seasonal drift should be considered. If your application must face the sun or a well-lit back area, an artificial front light may improve your view. Even in interior applications, a camera focused looking out on an entrance may be affected by outside light.

Cameras can be color or black and white. Certain applications may require color, such as monitoring an access control point where it would be necessary to see color-coded identifications. Black and white generally has sharper resolution, and may accomplish your objective better than color. Color also requires a light source with high color rendition. Placing a color camera outdoors with low-pressure sodium will not render proper lighting to distinguish colors.

Camera lenses come in different shapes and sizes. Lenses can be a prime lens, also called a fixed lens. This lens has a fixed focus and is aimed on a particular spot, called the focal point. Objects in front of and behind the focal point will be out of focus, the resolution decreasing the further from the focal point. This decrease is called depth of field, which can be improved by the size of the camera aperture; the smaller the aperture, the better the depth of field. Technology has improved this problem with autofocus lenses that adjust to the object rather than a fixed point. Fixed lenses come in different sizes depending on the location and field of view needed for the application. The further the camera from the focal point, the less the field of view. For example, a 12 mm lens with a 30° focus 10 feet from the focal point will have a 4-foot field of view, while a 6 mm lens with a 45° focus 10 feet from the focal point will have an 8-foot field of view. There are calculators to help determine the lens size and distance requirements, but it is better to define your needs and the contractor will select the right lens and focal point for the job. If you contract for a specific lens and distance camera, and it doesn't do what you want, it is your cost to make it right. Telling the contractor what you want and making corrections to meet your requirement is the contractor's cost to make it right.

A lens can be wide angle, capturing an image wider than normal, to long focus, capturing an image at great distances. The zoom lens combines the two features. Zoom lenses have a varied focal point that allows for one lens to "zoom in" or "zoom out" on a point and be in focus. Zoom lenses generally produce a lower-quality image than fixed focus. While technology has improved this degradation, keep in mind your primary objective in fitting a camera to its purpose. Add to this a pan/tilt mounting feature and the camera can be directed to different areas as needed.

Camera size can be very small and produce a unit that can be installed covertly. This may be helpful in applications where you do not want people to know they are being watched, making covert operations useful in internal investigations. Security applications are proactive and preventive, focusing on keeping the event from occurring, not catching the person in the act. The presence of a visible camera or even signage stating cameras are in use, is often enough to drive the wrongdoer away.

While the priority of placing and positioning cameras is observing the right place and thing, placement to avoid attack or compromise is important. You will not be able to prevent all compromise, but careful consideration of camera placement will minimize the opportunity. If the unit is too close to the ground, attackers can physically move, remove, or break the camera. Less damaging, but equally effective methods of attack are placing tape across the lens, placing a bag over the lens, and using paint to cover the lens to incapacitate a camera. Using a paintball gun is one way to reach even high mounted cameras. Mounting a camera on a wall where unauthorized persons have roof access, or at least some access above the camera, subjects the unit to attack by dropping something on the unit to disable it. Exposed cables or power lines can be disconnected or cut. Placing the camera on a pole or other extended mount decreases the success in disabling the unit. Camera housings protect the units from expensive repairs due to vandalism. A plastic cover is less costly than a camera and lens. The cover also protects the unit from weather. Covers are clear or tinted. Tinted domes decrease glare and prevent attackers from seeing which way the camera is pointed. Units can be heated for cold weather with humidity control and lighted for specific applications. Some units have a fan to control fogging. Bomb-resistant units are available. Camera power can come from a hard line power supply or from solar and battery power in hard-to-reach places.

Getting the camera image to a monitor requires a transmission method. The most common is hard line using Ethernet cable, fiber optic cable, coax, and wireless technology. The system can be networked with an IP address and accessed through Internet connections. This makes monitoring from multiple locations possible. For example, in a residential application, the homeowner can access his or her home system for monitoring a babysitter, receive alarm alerts and view the affected area, and even monitor events from a smart phone. With Internet access, Internet security and bandwidth are additional points to factor in your application.

Monitors should be color capable and of sufficient size to meet your needs (detection, recognition, identification). Multiple cameras can be split and observed on a single screen. Many applications will have a primary monitor viewing one camera at a time and other monitors in split-screen mode monitoring several cameras at the same time. The operator has the ability to direct which camera is on the primary monitor as well as controlling pan/tilt/zoom camera functions. Older systems use dedicated monitors that are designed with high resolution.

Cameras may or may not record video. Your objective may not require the recording of video, but video provides a record of events which may be useful in legal proceedings, investigations, or business decision making. Computerized systems and digital storage make it less expensive to record video. Systems can be programmed to record at a lower frame rate when there is no motion. When motion is detected, the system records the motion in a faster frame rate. Recording is real time; framerate is just how the computer will store the images. For example, no motion may be recorded at 2 frames per second and motion recorded at 30 frames per second. This allows for more storage space and efficiency in locating past events. Older units may record on VHS or DVD. This technology is quickly being outdated, but is still used in security applications. VHS and DVD have limitations on recording and storage capability. If you are operating with one of these systems, careful planning will maximize your viewing and recording capabilities. Keeping video depends on need and usage. Unique events can be culled and stored individually. Otherwise, digital storage will write over previously recorded images in a loop fashion, recording over the oldest data first. VHS and DVD can also be reused.

Cameras should be checked periodically to assure positioning is capturing the target area. Intentional or inadvertent moving of the camera may leave blind spots. Pan/tilt/zoom cameras are often put in a "patrol" mode where they scan back and forth. The pattern should vary so potential threats cannot determine the program. Also, pan/tilt/zoom should have a default setting to return the camera to normal operations after being manipulated by personnel. For example, a security officer may move the camera to watch an event, and forget to return the camera back to patrol. A default setting will automatically return the camera to normal operations.

CCTV surveillance use is extending beyond normal proactive security operations by utilizing "intelligent" video. Digital systems can be integrated with alarms to position cameras when an event is triggered. For example, a college programmed its security pull stations to CCTV monitors. When a station is activated, cameras move to their preprogrammed position to observe the area of the pull station. Cameras with no view of the station move to cover possible areas an attacker may use for escape, thus providing a better opportunity of identifying and apprehending the violator.

A simple system using CCTV for intrusion detection is for the system to recognize a change in pixel color/shade and sound an alert. The camera should be presented on the main monitor for security officials to assess the alert and determine if further action is necessary. To minimize false alarms, such as a bird flying through the field of view, users can define specific areas in the camera field of view for the alert. Alarms can be set for different activities. A static object alarm alerts when a stationary object remains in an area, such as an automobile parked on a bridge. Moving vehicles would not set off an alarm, or the program can be set to alert authorities to a change in pattern movement, such as a collision on the roadway slowing the movement of traffic. An object removal alarm would set off an alarm when an object from a defined area is moved out of the area. A path detection alarm alerts when movement of an object is in a particular direction by using multiple defined areas and following the sequence of movement from one area to an adjoining area. Path direction can be set to alarm when the movement is in one direction only. For example, an object the size of a person moving away from a protected area would not sound an alarm, but would if the object is moving toward a protected area. Areas can be blocked so you will not trigger a false alarm. For example, watching a doorway in a hall may pick up other hall traffic. The defined area can be reduced to cover just the doorway. You can also define other areas for the system to ignore, such as movement through a window next to the door.

Biometrics is being used with video capture as a nonintrusive means of identifying persons. This technology measures points in facial structure of a known subject. A CCTV image is scanned for faces and compares the unknown faces to the database. When an image is matched, it alerts the user. Depending on the purpose of the comparison and match, the user can take appropriate action. For example, a workplace photographs new employees for ID badges. The biometric system scans the photo and stores it in the database. If an employee is dismissed and tries to reenter the facility, facial recognition technology would alert personnel to an unauthorized entry. In contrast, as employees enter the workplace, CCTV cameras scan the faces for matches. When a face matches one in the database, the employee continues to a checkpoint. If a face does not match the database, an alert notifies personnel of a possible unauthorized entry, or employees can be notified of a VIP arriving or a problem customer.

The technology is highly accurate, but not without error. Biometrics are highly accurate, with many companies advertising their system can detect a bearded person who has shaved. Systems can also account for pose variations, skin color change, eyeglasses, and even facial expression. The initial alert is nonintrusive; the person is usually not aware he or she is being scanned. Discrepancies justify security personnel approaching the person for further investigation. Research is working to compare behavioral image patterns and alert authorities when a person exhibits certain characteristics.

14

Response Force

BOBBY E. RICKS

Contents

Security response forces are part of a comprehensive security program. Facilities that operate without a trained response force must rely on local law enforcement to respond to an event with an initial response by management who may or may not have any knowledge or foresight of security procedures or needs. A receptionist or clerical employee assumes the task of screening visitors and the initial response to a hazardous event. Distractions from his or her normal tasks, and the fact that security is not his or her forte, means relying on such for security is minimal, if nonexistent. The decision to have an on-site response force takes into consideration the volume of employee and visitor traffic, the value or sensitivity of the assets and resources being protected, local crime rates, the availability of local law enforcement to respond, and their response time, and other security measures in place.

A security force is a deterrent to criminal activity. Uniformed officers send a message that the site is guarded and an immediate response is available. Officers can supervise access control. They can secure exterior areas by having a guardhouse located away from buildings to verify authority for persons to be on the premises. Employees and visitors can be screened, and deliveries checked before moving on to buildings. At building entrances, guards verify the identity of employees whose access control badge is misread by electronic systems, as well as operating screening equipment such as magnetometers and x-ray equipment. Extremely secure sites may have explosives sniffers and the capability to x-ray vehicles as well as vehicle barricades.

Visitor badges can be issued and controlled, as well as having officers to escort key visitors to the site. The force works with human resources to issue identification badges or cards, entering information for electronic access control. They input new employees into the system and delete and reissue lost badges. In addition to badge and ID control, security monitors of stationary posts are needed for access control, and monitoring of intrusion detection systems, to include alarms and closed-circuit television monitors. While not necessarily a security function, security is used to identify and monitor hazardous conditions such as a chemical spill or other accident.

Security patrols can be used to respond to alarms, fire, or other security or safety concerns. Patrols can be dispatched to escort fire or emergency personnel to a location. Security barriers may need moving to allow for equipment to respond, for example, one site closed off the drive adjacent to the building for security purposes. Portable jersey barriers were placed at the ends of the drive to prevent vehicles from entering. In the event of a fire, the security personnel brought a forklift to move the barriers so the fire equipment would have closer access to the building.

Patrols also monitor for open doors and unattended equipment. During patrol, guards can identify and record security concerns such as broken windows or locks, or safety concerns such as loose steps, hallways lights out, or emergency exits blocked. More intensive patrols can check for open files, computers left on, and other issues mandated by security policy.

Patrols can be monitored by tour wand systems that record when a guard has patrolled the area by having an electronic wand touch a memory button fixed to a wall or door at the facility. A patrol tour would be set by placing memory buttons in strategic places that assure the patrol would pass by and view specific areas on his or her patrol. A wand is used to record the patrol. Managers download the data and can assure that patrols are being performed. An example would be a site with multiple buildings and perimeter fencing. Policy may dictate that security would patrol the site hourly. Memory buttons may be placed in and outside of buildings, gates, and fences to monitor patrol progress.

Determining the manpower requirements to operate a private response force requires a detailed analysis of the security response requirements. Initial considerations are how many stationary and patrol posts are needed and whether the post will be operational 24 hours a day, 7 days a week, or operational during certain times.

How many stationary posts are required? Are these posts manned 24 hours a day or at select times? For example, an employee entrance may require one guard present 24 hours a day, but at high demand times like shift changes, other guards may be needed. An employee gate may be open during high demand times, and then closed. A patrol guard can work the gate for the limited period it is open, and then go back on patrol. Delivery gates may be opened for a full day, but not at night, necessitating a guard for a full shift. The security office, with the alarm and closed-circuit television (CCTV) monitors, can be strategically placed so the guard can monitor the controls and assist in other functions during slack hours. The security manager should consider that applying one person to multiple jobs decreases the efficiency of that person. A guard called to monitor cameras while manning the front desk will have his attention diverted when people need assistance. During that time, the monitors will not be observed.

How many patrols are required? Security policy or deterrent needs help determine the number of patrols needed. A large facility requiring hourly building checks will need more patrols than a small site with few buildings. Patrols can be adjusted for time like stationary posts. For example, a 24-hour patrol post may fill in at a gate during a 2-hour shift change time. Patrols also provide relief for stationary posts for breaks. A supervisor generally assists on patrol and reliefs as necessary. Because of unanticipated events, the supervisor should not be assigned to a regular patrol. How will the patrols operate? Will the guard be on foot, in a vehicle, or both? Are there specialty vehicles guards will use, such as a golf cart or bicycle?

How many people do you need to fill these posts? One post, operational 24 hours a day, 7 days a week, requires 4.2 persons. The security manager must consider this as well as scheduling to justify the manpower needed to fill the posts. While most work will be done by full-time personnel, part-time people may be used during certain times or days. For example, a site calling for one guard may employ four guards, one of whom is the supervisor, working four 40-hour shifts, and one part-time guard working one 8-hour shift. Days off, sick days, and vacation time must be considered. In extreme circumstances, guards may be required to work overtime. Scheduling overtime increases manpower costs. Employee efficiency decreases with time, so managers need to be aware of excess overtime. Instead of having a guard work a double shift to cover for a sick coworker, it may be advisable to have one guard work over for one-half of a shift, and another guard come in one-half a shift early. Another consideration is time needed for training or administrative matters.

Shift changes for the guard force should not conflict with the workforce. If the workforce arrives at 7:30 for an 8:00 start time, you don't want a shift change at 8:00. You may decide that a 6:00 to 2:00 shift is best, so you have fresh guards on as people arrive for work and there is no confusion as to briefing the oncoming shift of special needs and circumstances while the urgency of personnel ingress is happening. A fresh guard should be rested and ready to work. He or she will not be hasty to "get this over with and go home."

What equipment will the guards need? A guard force should be uniform, meaning anyone can identify a guard for crime deterrent or in the event of an emergency. There are normal needs for a guard force, such as uniform shirts and pants, cold weather coats, and raingear. A hat is customary, but not necessary. Most uniforms are military/police style to present a show of authority, but some response forces use a less aggressive uniform look, such as slacks and a blazer. Badges also display authority. People joking about security guards often refer to "sew-on badges." Metal badges have a show of authority and gain more respect for the guard.

Safety gear should be provided. If a facility has hazardous chemicals, do the guards have protective equipment in the event of a spill or leak? Steel-toe shoes or hard hats may be required in certain areas. Radios, tour wands, and flashlights may fill out the complement of equipment needed for duty. If vehicles are used, the security manager should assure the guard is capable of operating the vehicle and is a properly licensed driver.

While guards may be required to provide some of their own equipment, policy should dictate uniformity. For example, most guards will wear black shoes or boots. The type of shoe may be left to the user with limited parameters (e.g., shoes must be black), or severely restricted, specifying the type of shoe allowed (black patent leather with steel toes).

The organization should develop standard operating procedures for the guard force to follow. The duties and responsibilities of the guard force, as well as individual posts, should be identified and published in a procedures manual. See Chapter 8 on policy and procedures for more information.

The decision to provide defensive equipment should be determined by policy. Consideration as to nonlethal force should be given the same care and concern as deadly force. If the decision is made to allow weapons, training should be provided to assure proper use. If the decision is made not to allow weapons, policy and procedures should dictate the protocol for the protection of the guard, e.g., call for the police rather than attempt to detain a person. Along those lines, policy should detail the authority, if any, for the guard force. Some jurisdictions give limited authority for guards to detain. In other situations, citizen's arrest authority or the decision for guards not to arrest or detain should be clearly stated in the policy.

Now that you have determined the personnel and resource needs for your facility, you may choose to have a proprietary or contract guard force. A proprietary force is owned and managed by your organization. They are your guards, and your organization is responsible for hiring, salary and benefits, and training and equipping of the guards. Liability rests solely on your organization for intentional or negligent acts of the individual guards. A contract guard force is managed by a contract company (often a company specializing in guards) for a specified contract price. They hire, provide salary and benefits, and train and equip the guards. Your organization will have someone who will be the liaison with the contract agency. In some matters, the guard company may be liable for intentional or negligent acts of a guard. The liaison will be responsible for assuring the contract agency is performing according to the specifications in the contract and meeting with agency representatives for routine performance reviews.

Guards may be represented by a union and operate under a collective bargaining agreement with management. If you have a proprietary force, that is you. The security manager must work with human resources and legal counsel to assure that discipline, scheduling, and promotions are consistent with the agreement.

Hiring the right people requires following a process that assures you get the people you want without violating a person's rights. The objective is to hire responsible, qualified people that can and will work with others. There are legal requirements to follow in the hiring process. Title VII of the Civil Rights Act of 1964, the Equal

Employment Opportunity Act of 1972, and the Americans with Disabilities Act assure that people are hired according to their ability and not because of real or perceived disabilities or prejudices. Work with the human resources department to assure you follow all legal guidelines when hiring. Focus on whether the person can perform the job. A job task analysis will determine the essential skills, knowledge, and abilities a guard will need to perform his or her duties—things like walking, lifting, driving, sitting, standing, communication skills, both written and oral, reading skill, and vision. This is just a start. You may identify other requirements to meet your needs, such as a security clearance. State or federal law may dictate job requirements. For example, a convicted felon could not be hired to a post requiring the use of a firearm.

Screening and assessment involves many possibilities, such as work assessments, testing, interviewing, physical fitness test, physical exam by medical personnel, or a background investigation. For efficiency, the screening process should go from the least expensive method to the most expensive.

When screening applicants, the application needs to include the information needed to complete the first assessment. For example, if you require a high school diploma, the application should have a place for the applicant to list education. If applicants need to submit additional information, list the requirements in the job announcement. For example, you may require applicants to submit a writing sample. As you screen out unqualified candidates (e.g., under 18, no diploma), the remaining applicants can be evaluated through a variety of means: testing, skills use, etc. This process may eliminate other candidates and also begins to rank candidates in terms of skill, knowledge, and abilities. For example, a writing assessment may disqualify an applicant for failing to meet a minimum standard, while at the same time rating other applicants as good or excellent.

Another part of screening is to interview the applicant. Behavioral questions that focus on past experiences where the applicant has had a similar situation best reveal how the applicant is likely to act.

Regardless of whether you have a proprietary or contract force, a job assessment will identify the needs for the guard force as well as begin the process for writing a job description. To write a valid job description, begin with listing all of the tasks a guard will perform, such as answering the telephone, writing reports, watching monitors, and operating machinery. Any movements should be documented to justify requirements of the Americans with Disabilities Act (ADA). The ADA protects against discrimination of a person who can perform the job functions, with or without reasonable accommodation. For example, list tasks such as standing, walking, negotiating stairs, climbing, crawling, reaching, etc. Vision and hearing requirements, as well as the ability to operate specific machinery, should be included in the list. As you screen applications for employment, those who cannot perform the functions, with or without reasonable accommodation, can be excluded from consideration. Is it important to write a thorough job description so potential applicants will know the requirements and apply for jobs for which they are qualified.

The job description is also key to training. Training assures the employee can perform job tasks proficiently. It minimizes civil damages in the event of a lawsuit. Training may be in the classroom, online, on the job (OJT), or a combination of methods. Contract agencies provide a level of training for uniformity of their guard force. A proprietary guard force may be trained on site or sent to a school that provides a basic level of training. Private corporations and security organizations have ready-made materials for use in general training of guard forces. Depending on specific needs, other prepared training programs may be helpful and even necessary. For example, a guard at a casino where alcohol is served may have required training from the state gaming commission as well as the alcohol beverage commission.

Guards should be aware of the policies and procedures, company policies, operational needs, and priorities. Orientation should include a copy of the policy manual. After a reasonable time, have the guard sign a form indicating he or she has read the policy manual.

Training should include general duties applicable to the guard force as well as specific duties for each post the individual guard may work. Administrative matters such as paydays, insurance, and time off can be disseminated at this time. Additional training should be conducted based on the frequency or criticality of a specific function. For example, reports completed daily have a high frequency, so training on report writing would be a reasonable training need. Use of force may not be frequent, but highly critical, as the improper use of force can lead to injury and possibly lawsuits. Training the parameters of the use of force will assure proper conduct of the guards and mitigate damages if and when force is used.

Other training needs would be patrol or station procedures, customer service, telephone communications, and other skills not necessarily related specifically to the trade, but for events that could impact the organization. For example, when a guard answers the telephone or gives instructions to visitors, he or she must sound (and appear) professional. Answering the phone with "Yeah, what can I do for you?" is quite informal and may present a negative impression of the organization. Proper procedures should be identified and taught to all personnel.

Training should include all equipment the guard will use. This includes vehicles, patrol wands, weapons, and safety equipment. Guards are often the first responders to a fire, safety, or medical emergency. Knowledge and use of fire extinguishers and other firefighting equipment is essential. The guard should know when and how to deploy and use safety equipment, such as personal protective equipment, a fall protection safety harness, or safety barriers. Training should include Occupational Safety and Health Administration (OSHA) rules that apply to the facility and specifically to activities the guard may encounter and areas where the guard must patrol. Guards should be familiar with any first aid equipment and automated external defibrillators (AEDs), if available. Guards should have a working knowledge of emergency evacuation and shelter-in-place procedures. The guard force should always participate in all drills and exercises. Most organizations expect the guard force to coordinate emergency activities.

Leadership and the Response Force

Supervision and leading the response force to provide professional, efficient, and effective protection is the role of the leader. Supervisors often "supervise" with little or no training, relying on what they have been exposed to in past experiences. Leadership motivates people. People design, develop, and implement work processes. Leadership motivates people to follow and achieve process goals. Management controls systems by measuring the processes to achieve an efficient and effective flow of resources for productivity. Like management, leadership can be learned. Anyone has the potential to be a leader.

Situational leadership is a model focused on directive and supportive behaviors. Depending on the situation, the leader can change his or her style for an ideal fit to a follower's needs. The leader assesses the follower's competency and commitment to determine what style of direction is needed. The leader assesses the employee's development level and selects a leadership/supervisory style to match that level. Situational leadership has survived the test of time and proven to be an effective means of training leaders within an organization. The system works by having the leader apply the proper style to the individual follower. It is easy to understand and apply.

Take the example of a new employee. He or she is committed to do good work, but may not have the level of skill and knowledge to perform the tasks. The supervisor will be effective in directing the behavior of the new employee. As the employee develops competence, the supervisor moves to coaching, continuing to give direction and adding high support to the process. Eventually, the employee develops high competence and the leader can delegate tasks totally to the employee.

Contingency theory is similar to situational leadership in that leader success depends on the leader's ability to match his or her style with the situation. Contingency theory is supported with research that validates its ability to explain how effective leadership can be achieved. Like situational leadership, contingency theory is not tied to one "best" way of handling a situation, but rather teaches the leader to be flexible in his or her approach. Contingency theory sees leaders as either task motivated or relationship motivated.

The team leadership model integrates mediation and monitoring concepts to a group rather than to an individual. Effective team performance begins with a mental model of the situation. The team leader develops a mental picture of the situation and relates it to the team. The team then takes action to solve the problem. The team leader must decide to intervene in the team's efforts by asking three questions:

1. *Should I monitor the team or take action?* The leader watches for internal or external factors that need attention, and then must decide if his or her input is needed.
2. *Should I intervene to meet task or relational needs?* When the answer to question 1 is to take action, the leader focuses on

the need to be addressed: Is there a problem with the task, a problem with group relations, or both?

3. *Should I intervene internally or externally?* Internal support is needed at times when you must clarify goals, facilitate decision making, and emphasize standards of excellence. External support is when you must work in the organization to give the team credibility. External support may be gathering information from other functions so the team can function efficiently.[1]

Endnotes

1. Northouse, P. 2010. *Leadership: Theory and Practice*, 5th ed. Thousand Oaks, CA: Sage.

15

Information Technology and Security

TRUETT GRAHAM RICKS

Contents

With the prevalence of computers in the workplace, and businesses integrating more and more technical systems, dependence has been created on computers and technical systems. This dependence makes computers and technical systems an easy target for those looking to gain access to a business's data. With businesses storing the majority of data on systems with access to the Internet, also meaning there is access from the Internet to those systems, it is critical to do everything possible to protect that data. In recent times, data breaches such as those of major retailers resulting in theft of the data of millions of individuals, and even data breaches of government sources leaking data have served to show the importance of data and information security. Such data breaches not only expose sensitive information, but also can damage relationships with clients whose data may be exposed.

The level of security required by any network is wholly dependent on the type of data contained on the network. There is a large difference in the necessary level of protection on a network for a law office than there would be for a florist. When analyzing the information security of a business, the audit is typically in three parts: workstations, network equipment, and people. These three areas are essential to the security of technical systems. This chapter will look at these areas and the most important elements of each for a business owner or manager to be aware of.

Workstations

When performing a security audit of a business, the first on-site task is to check the workstations. The security of workstations is the most basic level of security, one that any business should consider important. In the case of an attack on a business, the workstations are the most common target. In a business, workstations are generally plentiful, and they tend to be easy to physically access. They also have a few inherent weaknesses that make them easy targets for those looking to gain unauthorized access to a network. There are free tools readily available online which allow someone with a very basic knowledge of computers to wipe a password for a local user account in a few short minutes. This is an act that allows unrestricted access to the computer. One very important thing to consider is that an attack on a local computer is not limited to someone sitting down at the computer and using it. An act as simple as inserting a hardware keystroke logger between the keyboard and computer or even a simple USB flash drive inserted in the back of a computer can transfer information or give access to unauthorized persons.

When checking the security of a workstation, it is important to begin with the most obvious and easy-to-correct issues first. The task is not just closing all of the holes, but closing as many as fast as you can. If closing one security hole takes an hour of work but there are five other holes that can be plugged in the same amount of time, always start by closing the five holes. It is important to remember there is a careful balance that must take place between closing big holes and closing multiple holes; for this reason, most technicians will work from easiest to hardest to close.

Software

One of the easiest security issues to correct are ones related to software. The most obvious software-related issue is also one of the most common poorly executed ones: antivirus and antimalware protection. When it comes to having antivirus and antimalware protection on a computer, there is a balancing act that must take place. While it is important to have protection, there is a point at which protection becomes a problem. One factor to understand is there is a difference between antivirus protection and software that is intended to protect against malware and spyware. Just because you have one type does not mean you are protected against the other. A software solution to protect against viruses is the most important type of protection to have, but it is advantageous to supplement it with software providing malware protection.

When looking at antivirus solutions, there are a few important items to look for. A good antivirus should not only provide real-time scanning and frequent definition updates, but also be capable of scanning inside of compressed files, such as .ZIP files, and scanning email attachments. Real-time scanning is an important feature because the software is constantly scanning files looking for infections. Software without real-time scanning requires scans to be run by users or system administrators on a regular basis. Virus definitions are the files that tell the antivirus program how to detect viruses, and they require frequent updates to make sure they are able to detect new viruses. If a software solution does not frequently update virus definitions, then it is less likely to be able to quickly detect new viruses. Being able to scan for viruses inside of compressed files or email attachments is very important, as those are both very common methods for transmitting viruses. Any software used to protect against malware should have similar features, but with real-time scanning being less important, in such a case the software should have the ability to schedule automated scans.

The topic of viruses and malware brings up a very important point: all computers are vulnerable to viruses. Viruses infect any system that is designed to run applications, and a virus is nothing more than an application with a malicious intent. Viruses are a less common occurrence on Macintosh and Linux-based devices because they hold a smaller portion of the computer market. A person writing a virus is more likely to write a virus that can target a larger number of computers. Apple used to list being virus-free as a benefit of its computers; in recent years those statements have been removed from its website (Mlot, 2012). These days most of the large antivirus software companies have versions written specifically for Macintosh-based computers.

Another software-related security issue is the installation of unnecessary software on a workstation. Software companies are constantly releasing updates for software they make, not only to improve how the software functions, but also to fix security holes and issues. Removing unnecessary software reduces the chances of a software not being updated or having a vulnerable security hole. This is why it is absolutely critical to make sure that all software on every machine is always on the latest version. This also includes making sure that the workstation has every applicable update from the creator of the operating system, whether it be Microsoft, Macintosh, or Linux. Some of the more common types of unnecessary software on a computer come in the form of toolbars, free PC optimizers, and free games. These types of software should not be installed on computers and frequently lead to virus infections by creating additional holes in a computer's security. The installation of unnecessary software by users is an easily addressed issue; the easiest fix for this is to make sure that standard users do not have administrator rights on the computer. In the majority of cases it is not necessary, and it is a best practice to restrict installation rights for software from standard users. This not only is a security issue, but also prevents the installation of software that detracts from productivity, such as games.

Correcting security holes related to software is generally simple and is one of the first areas tackled by technicians. One critical point to understand is that these software issues are not limited to workstations, but also affect servers and, in some cases, mobile devices such as smart phones and tablets. Software-related attacks, whether from viruses or even software that is unknowingly malicious but intentionally installed, are some of the favorite methods for gaining access to a workstation or network. This is because the person gaining access may be on the other side of the world and nearly impossible to trace. These attacks are easy to perpetrate because they can be sent through email attachments to tens of thousands of people at once, or a piece of software, seemingly harmless, posted on a website one time where millions of people could potentially download it.

Hardware

After analyzing the state of the software on the server, the next thing to consider is the hardware security of the network. The physical security of computers and a network is commonly overlooked, but it is just as critical as having antivirus protection. All of the protection against software vulnerabilities means nothing when the computer is stolen. In such a case the person committing the act has all the time he or she needs to get the data and not be adversely affected by the software protection.

One method that is helpful in mitigating the risk of data theft from the physical theft of a computer is the full encryption of the machine. Full disk encryption means every sector of the hard drive inside of the computer is encrypted. This requires that upon starting up the machine, a password be put into the machine that then allows access to the rest of the drive, so that the computer can boot. Without full disk encryption, if a computer or even just the hard drive is taken from a computer, the drive can simply be put into another computer as a secondary drive, and all of the data accessed without trouble. With full disk encryption, even if the drive is taken from the machine, the data is still fully encrypted and not accessible without the encryption password. Full disk encryption should be considered necessary for any machine that may store important or confidential data. It should also be considered necessary for all laptops. Laptops are a very common target for theft, even if the target is not the data themselves, whether from inside of a car or even while it is sitting right beside the owner in a busy airport.

Locks are one of the oldest methods of security, and computers also benefit from locks. Computers and even monitors have for several years been equipped with Kensington Security Slots. These slots enable the connection of cable-type locks similar to bicycle locks. While they tend to be easily defeated by a pair of wire cutters, they do provide a deterrent for quick smash-and-grab break-ins. Besides Kensington-type locks, most desktop computers now contain a special place on the back of the computer where a standard combination or key lock can be placed to prevent the computer from being opened. While this does not help in the case of a computer being taken from the location, it does keep the case from being entered while in the office. Most new computers utilize toolless designs, which allows installation or removal of components quickly and easily. In such a situation, a computer can quickly and easily be opened and the hard drive removed in less than a minute, leaving the rest of the computer behind. This allows a thief to take the part of the computer containing all of the data without anyone noticing until he or she tries to use the computer. Similarly most servers have locks built in to the front of the case that prevent access, and enclosed server racks generally have locks built in. While these locks are generally small and can be defeated with simple tools, they provide an additional barrier to theft and unauthorized access.

The location and access to the primary network equipment, such as servers, switches, and routers, is another important factor in security. Generally, these devices should be kept in their own room, and that room should be centrally located in the building away from exterior walls and have a locked door. It is critical, though, that when kept in a locked room, the room is cooled. It is common for businesses to put such equipment in a utility closet that may also contain air conditioning equipment or water heaters. This is highly discouraged as it is risky to put critical equipment that controls your business and network in such close proximity to water. Through all of this, it is important to keep in mind the goal is preventing unauthorized access. If a person looking to initiate an attack or steal data from the network can gain access to a room with networking equipment, they are essentially being given free rein of the network. Aaron Swartz, a self-proclaimed Internet activist that believed in the freedom of information, systematically downloaded a large number of academic journals from JSTOR on the campus of the Massachusetts Institute of Technology by plugging in a laptop to a network switch he found in an unlocked network closet. The only time he spent in the closet was to initially plug in the laptop and subsequently change out external hard drives with data on them. He was caught as a result of a video captured from a security camera placed in the closet after a change in network traffic had been noticed. All that was required was to lock the door to the closet.

Network

Network security is very simple: keep unauthorized people out. When securing a network, just like the workstations, you have to close all of the holes. Having highly secured wireless, servers, and general network is worthless if the router is insecure. One of the most important aspects of network security is the router itself.

A router is the network device that controls the flow of data across the network. It manages the network, including traffic entering and leaving your network. For a business it is important to ensure that the router is business class. Only a business class router can provide the level of security and management required by a business. If your router is set up with a wizard, it is most likely not business class. Business class routers will cost more than a consumer-grade router intended for use in a home, but are required to provide adequate management and security. Consumer-grade routers tend to focus on speed and easily configured security. Business-grade routers put focus on speed as well as full-featured security and remote access.

A quality business class router will allow you to have full control over your network. It will also have features such as virtual local area network (VLAN) management, which allows for the network to be broken into separate networks to further control and restrict access to parts of the network. One example of this is having a guest wireless on its own VLAN so that it is completely separate from the rest of the network, allowing Internet access but not access to other devices on the network. Another feature of quality business class routers is support for multiple virtual private networks (VPNs). VPNs allow remote locations and devices to connect to the local network through an encrypted tunnel, allowing users to utilize the same network resources as on-site users. These two business class features are critical to the security of the network.

In addition to business class routers, managed switches are important in securing a network. A switch is the device that allows multiple devices to connect to the network through an Ethernet connection. A router may only have a few connections intended for the local network, so a switch is used to take one of those connections and turn it into 24, 48, or any number of ports if using linkable switches. A managed switch allows for complete control of each of those ports. It allows not only for seeing how much traffic is passing through each port on the switch, but also for security features such as VLANs and even disabling ports that are not in use.

Disable unused network ports. As previously mentioned, the securing of rooms containing networking equipment is critical for security, but if there are active network ports in parts of the building not closely monitored or public areas, they should be disabled. These active ports are an easy point of access for someone looking to attack a network.

A server is an important aspect of every network, and while servers can be very complicated, securing them is simply a matter of making sure a few things are correct. If the rest of your network is secure, it makes securing a server that much easier. The most important security item, when it comes to a server, is to make sure people only have access to what they need. This not only makes management easier, but also makes sure that a low-level employee does not have access to the same documents and resources as high-level employees. Utilizing a server role called Active Directory, individual user accounts are set up to allow customized permissions for each user. Another feature of Active Directory is that it allows workstations to be added to a network feature called a domain. This allows any user to log on to any machine with his or her Active Directory account. This allows flexibility in which users use which workstations while not allowing access to other users' documents on the workstation.

Active Directory also allows administrators to control user accounts in ways such as creating complex password requirements and even monitoring users' behavior on the network, such as what files they are accessing. A complex password is generally considered to be at least eight characters with a mixture of uppercase and lowercase letters, numbers, and special characters. This makes a password hard to crack with brute-force methods, which literally mean starting with *a* and trying every combination of numbers, letters, and characters. It is recommended that password policies should require passwords to be changed every 90 days and not allow a user to use any of his or her last four passwords. It is also recommended that a password should not contain any words or have any special meaning to the user. These are commonly used in order to make a password easier to remember, but they also make them easier to crack or discover through social engineering. The most important aspect of passwords to remember is passwords should *never* be written down.

In many cases a business claiming to have a server hosting its data has nothing more than a workstation with all of its data shared off of it. This is not a viable server and a poor choice if security is a concern. A server is a higher-quality machine that is running a server operating system. The most common examples of server operating systems found in businesses today are Microsoft's Server 2008, Server 2008 r2, and now Microsoft's Server 2012 and Server 2012 r2. Having a server operating system means that many extra security measures are in place and there are restrictions on the server to help avoid things such as attacks on the server and viruses. Just like a workstation, though, a server should have an antivirus software solution in place to help provide an additional level of protection. Servers also require updates, just as workstations do, so both the operating system and installed software must be updated regularly.

If a server hosts any form of data or applications used by a business, then it is important to make sure that access to that server is restricted from outside of the network. This concern comes into play mainly when a business is hosting its website on a server that is physically on its local network. Most businesses utilize web hosting from third-party providers, but some businesses still host their own sites. If this is the case, it is critical to understand that people accessing the website are accessing the server. This is what is known as an Internet facing server. These web servers should never be the same server on which a company's data and applications are hosted. Doing so provides someone looking to gain access to the network a place to begin his or her attack.

Wireless networks are becoming commonplace in businesses, especially as wireless devices such as tablets and smart phones become more accepted as standard. Many businesses also offer wireless Internet access to clients and guests. As mentioned when talking about VLANs, if a guest wireless network is present, it is important to make sure it is separate from the rest of your network. An easy method for someone looking to gain access to the network is to simply sit in the parking lot of the business and attempt to connect to the wireless network. If a guest network is not separated from the rest of the network, then it is easier to access local network resources. Guest networks, if separate from the rest of the network, may have limited security generally in the form of a generic password to connect. Some new wireless routers allow the creation of a guest network that is separated from the rest of the network through configuration on the router itself. Routers that offer this functionality tend to lean more to the consumer side of the market, and generally the VLAN method of creating a guest wireless network is more reliable.

As wireless technology becomes more prominent and pushed into the business world, some businesses have taken to having a strictly wireless network. Wireless will always be less secure than a wired network, because of the nature of the way it broadcasts traffic even when secure. A nonsecure wireless network not separated from the work side of the network will allow an attacker the same level of access as sitting at a workstation inside of an office without ever having to enter the office. While there exist numerous methods for securing wireless networks, many security methods still offered by wireless routers and access points have been deemed highly vulnerable or already cracked. One previous standard for wireless security that is still offered on many routers and access points is Wired Equivalent Privacy (WEP). This form of security has been shown to be easily cracked with free tools and a wireless laptop with tutorials on cracking WEP are readily available online. At the current time, the lowest form of security that should be employed on any wireless router or access point is Wi-Fi Protect Access 2 (WPA2). As technology continues to evolve and people discover new vulnerabilities and holes in security, these wireless security standards are modified and updated. This is a reason to always employ the highest level of wireless security available and to keep up to date on the newest security versions. One of the newest methods of wireless security is called 802.11i, it is a variation of WPA2 with an additional level of encryption but is only supported by a very limited number of the newest wireless hardware.

A common method for securing wireless connections, which has been used for years, is called media access control (MAC) filtering. It is common to see businesses employing this method even though it has a very simple-to-execute attack that bypasses it. Every network device, whether wired or wireless, has a unique number assigned to it called a MAC address. MAC filtering is simply telling the wireless router which devices are allowed to connect by providing that unique number to the router. This allows the devices on the trusted list to automatically connect with no other action required by the user. The problem with this type of security is that a MAC address can be "spoofed," making a computer appear to have a different MAC address; this is easily done by anyone with a little networking knowledge or ability to perform a simple web search. The only thing required to execute this attack is a MAC address that has permission to access the wireless network. In most cases, even on secured wireless networks, the MAC address of a computer is broadcast in plain text repeatedly over the wireless signal, making it easy to intercept the MAC address of a computer with wireless access. This method is being mentioned only because of the extreme ease of bypassing it; any wireless network using this as its form of security is essentially not secure at all.

People

No matter what the type of security, the weakest link is always the same, the people. The most complex system of network security is useless if a user does not log out of a machine or gives his or her password to someone who should not have it. As a system administrator, there have been numerous times that I have called someone and asked for his or her password, only to have him or her give it to me without ever questioning who I was or what I was doing. Along the same lines, I have walked into the office of people who have never seen me, told them I was there to work on a computer, and physically taken the computer from their office without being questioned. I could have been anyone off of the street that walked in carrying a laptop and wearing a polo and stole one of their computers.

Social engineering and simple observation is one of the biggest threats to the human element of information security. People attempting to gain access to a network may employ a variety of methods to gain information, such as phishing emails or phone calls, or even simply looking at the information around the workstation. Many times a person may ask for information that may seem unimportant or completely unrelated but is used at a later time to complete a more complicated social engineering scam. One of the most basic techniques to gaining information that may allow access to a network is simply by looking around. Looking at things such as notes on desks, under keyboards, or even attached to monitors can reveal a lot of information useful to a person looking to gain access to your network and data. The only way to overcome this inherent security weakness is by training users. Users must be trained on not only types of attacks, but also how to deal with people claiming they need to work on the computers. Things such as checking an ID or even calling an office to make sure it did in fact send someone to work on a computer are a quick way to make sure the person is who he or she says he or she is and add a layer of protection. Some offices will even take steps such as having photographs of the people that work on the computer by the receptionist's desk to make sure personnel know who is authorized to work on the computers.

Special Considerations for Cloud Computing

Cloud computing and cloud-hosted services are becoming more and more utilized in business. While this is still a relatively new form of doing business, it is being utilized and does have implications for security. Since this area is constantly evolving and there are so many ways that cloud computing is employed, there are many different security concerns. Cloud computing is the same situation as if a user were at a remote location connecting back to the office. The connection should be secured, preferably encrypted, and access must require a username and password. The servers that are hosting the applications and data should meet the same security requirements for a server on-site. Cloud computing requires users to transmit data over the Internet, which may create a vulnerability. When data is being transmitted, even when encrypted, there is a chance of the data being intercepted. For the highest level of security, the data should not leave the possession of the business, which means it should stay on on-site servers and workstations.

Summary

Information security is a vast topic on which thousands of books have been written and continues to evolve. This chapter is by no means intended to provide a comprehensive view of information security. It comes from experience dealing with managers and business owners, the most important topics for them to know, and questions they ask. Below is a list of the most important factors from this chapter provided as a quick reference.

Software

- Antivirus software is necessary regardless of the type of workstation; it should be updated frequently and provide real-time scanning.
- Either the antivirus software or the email server should scan email attachments for viruses and malware.
- Workstations should be free of unnecessary software, including, but not limited to, toolbars, free PC optimizers, and games.
- Software should always be on the newest version and updates installed as soon as they become available.

Hardware

- Computers containing vital or sensitive data should utilize full disk encryption for a high level of security.
- Computer cases, server chassis, and network racks with locks on them, or a place to put a lock, should remain locked.
- Servers and network equipment should be kept in their own room that remains locked and is centrally located in the building.

Network

- Routers should be business class, not consumer grade.
- Use VLANs if you need to segment a network to control access to certain parts, such as guest wireless networks.
- VPNs are used to allow encrypted connections from remote locations to the local network.
- Physical access should always be restricted to the network; this includes unused network ports.

- Passwords should meet requirements for complexity, such as eight-character minimum with a mixture of uppercase and lowercase letters, numbers, and symbols.
- Passwords should *never* be written down.
- Wireless routers and access points should use a minimum of WPA2 security.

Through all of this, one of the most efficient things that can be done to prevent security issues is to train the users. Users have to be trained on social engineering and attempts to gain unauthorized access to the network. They need to be trained to recognize potential threats and know how to deal with them.

Endnotes

Mlot, S. (2012, June 25). *Apple Strips Mac OS X Website of Virus Statements.* Retrieved from PC Mag: http://www.pcmag.com/article2/0, 2817, 2406275,00.asp.

Appendix: On-Site Survey Checklist

Completed by: ____________________ Date: __________
Client name: ______________________
Contact person: ______________________
Phone number(s): ____
Location address: ______________________
City__________ State_____________ Zip______________

This survey is used to conduct risk assessment of specified client locations. The surveyor/survey team should have a copy of the presurvey data on arrival at the site. Observe the physical and environmental conditions and operational procedures at each location. Interview the appropriate client personnel to complete the survey form. Comment on any high-risk situations as well as recommended actions.

I. Type of Facility

Construction	Office park	High-rise	Office building
Hospital	Bank	Hotel/motel	Residential

Retail: List products: ________________________________
Distribution center: Products: _________________________
Government-owned/leased facility
Manufacturing: List products: ________________________
Other: Describe: ________________________

II. Presite Visit

1. What is the general purpose of site?
2. What are the usual business hours?
3. How many people have access to the site?
4. Is site open to the public?
5. Which hours and days represent high-activity use?
6. Which hours and days represent low-activity use?
7. Identify organizational assets:

   Personnel
   Information
   Equipment

8. Rank assets in priority order.
9. Is there a Sensitive Compartmented Information Facility on the property?
10. Are there critical or restricted areas?

    How is access to these areas controlled?

11. Who are the tenants on your property?
12. Who are the tenants of the adjoining properties?
13. Who were the prior tenants on your property?

    Did these tenants report or record any criminal activity?

14. Do you have a neighborhood crime statistics report?
15. Source of report.
16. What types of crimes are the most prevalent in the area?

    List by type, actual number, per capita.

17. Do you have a log of events occurring on the property? (criminal or noncriminal)?
    List by type, date of event.
18. Are hazardous materials kept on the property?
    Type: Chemical, explosive, nuclear, biological, incendiary location.
19. Is there a guard or response force on site? Provide a copy of the policy and orders.

20. Provide a copy of emergency plans: Bomb threat, workplace violence, fire, shelter in place, etc.
21. Insurance coverage: Fire, theft, personal injury, business loss, etc.
22. Provide a copy of the security policies, including employment screening procedures.
23. Has there been a community terrorist threat assessment?
24. What is the terrorist threat level for this facility?

List factors identifying vulnerability.

III. Site Visit—General Assessment

1. Assessment of neighboring properties:
   How far out did you look?
   How close are neighboring facilities?
   Threat?

2. General appearance of the facility:
   Setback from perimeter
   How close are buildings on adjoining properties?
   Vegetation close to buildings, obstructing view
   Visible deterrent: Fence, sensors, CCTV, etc.
   Clean, orderly, or unkempt
   Obvious security hazards
   Trash/dumpster location/appearance

3. Attitudes toward security:
   Any unlocked doors?
   Computers/documents unsecured?
   Display of badges/parking decals

4. Ease of access by emergency vehicles
5. Frequency of police patrol/response time
6. Distance to nearest fire station/response time
7. EMS response time
8. Any high-risk situations, unusual activity, or other issues observed?

IV. Security Policy and Programs

1. Has management established and communicated a security policy?
2. Have security procedures been published?

   Date of last review: ____

3. Are the employees frequently updated on security procedures?
4. Is there a designated individual to supervise the security program at this site?
5. Is there a file of security deficiencies and a schedule for corrective action?
6. Is there a policy for reporting security breaches?
7. Is there a policy for reporting criminal acts?
8. Is there a policy on criminal prosecution?
9. Does the company conduct preemployment background checks on applicants?
10. Does the company conduct preemployment drug screens on applicants?
11. Does the company conduct random or "for cause" drug screens?

V. Assets and Property Control

1. Is there an inventory control system?
   For equipment?
   For critical records?
   For retail stock?

2. Who is responsible for inventory control for each of the above?
3. How often are inventories audited?
4. Are specific procedures and documentation required for transferring items into and out of inventories?
5. Do employees require signed authorization to remove company property from the facility?
6. Are high-value items stored in a special area with additional physical security measures?
7. Is retail stock kept in a secure area?
8. Is access to retail stock controlled?

9. Is the retail stock area separate from other departments?
10. Is the retail stock area safe from physical damage, such as breakage, fire, weather, spoilage, etc.?
11. Have there been any thefts of property or inventory?
12. Are thefts reported to security?
13. Is there an investigation of thefts?
14. Are theft statistics compiled and tracked?
15. Are they reported to top management?
16. Are scrap materials promptly sold?
17. Are strict controls and accounting records kept for the sale of scrap material?
18. Is the sale of scrap or salvageable material a separate department and not handled by employees involved in purchasing, production, or sales?
19. What are the controls for cash control and accounting? (retail)

VI. Vehicle Control

General

1. Are parking lots provided?
2. Are parking facilities adequate?
3. Are separate parking areas provided for visitors' vehicles?
4. Are automobiles allowed to park so close to buildings or structures that they would be a fire hazard or obstruct firefighters?
5. Are automobiles allowed to be parked within operating or controlled areas?
6. Are interior parking areas located away from sensitive points?
7. Are interior parking areas fenced so that occupants of automobiles must pass through a pedestrian gate when entering or leaving the working area?
8. What is the extent of guard surveillance over interior parking areas?
9. Are there restrictions against employees entering private vehicle parking areas during duty hours?
10. Are automobiles permitted to be parked close to the controlled area fences?

Vehicle Registration

11. Have definite procedures been established for the registration of private cars, and are they issued in writing?
12. Are vehicles that are allowed regular access to the facility registered with security?
13. How often is registration updated? Annual?
14. What information is on the registration form?
15. Do the prescribed prerequisites for registration include a valid state registration for the vehicle, proof of insurance for the vehicle, and a valid operator's license for the registrant?
16. Do vehicle registration requirements include motor vehicles owned or operated by nonemployees whose business requires frequent access to the facility?
17. Does the regulatory control criteria for registration include:

    Prohibition against transfer of registration permit tags for use with a vehicle other than the one for which originally issued
    Replacement of lost permit tags
    Return of tags to the security officer when the vehicle is no longer authorized entry into the facility
    Destruction of invalidated decals or metal tags

18. Do registration permits bear a permanently affixed serial number and numerical designation of year of registration?
19. Is mechanical inspection of vehicles required as a prerequisite of authority to operate a vehicle within the facility?
20. Are decals or hang tags affixed to all vehicles authorized to operate within the facility?
21. Are temporary tags issued to visitors' vehicles?
22. Are there periodic checks to ensure that vehicles are operated on the premises only by properly licensed persons?

Company Vehicles

23. Is there a log of company vehicles?
24. Does the log indicate where vehicles are located? (parked on site or home fleet)
25. Does the custodian of vehicles have a copy of a valid driver's license for persons authorized to drive company vehicles?

26. How often must personnel update their driver's license information?
27. Where are vehicle keys stored? Are they secure?

VII. Shipping/Receiving and Controlled Traffic Areas

1. Is the shipping/receiving area within a protected area?
2. Is all vehicle traffic into the area controlled?
3. Are loading and unloading platforms located outside the operating areas separated one from the other and controlled by supervised entrances?
4. Do checks cover both incoming and outgoing vehicles?
5. Is the area under security surveillance? (manned or CCTV)
6. Do truck drivers or outside vendors have access to the shipping/receiving area?
7. Does the supervision of loading and unloading operations ensure that unauthorized goods or people do not enter or leave the installation?
8. Are truck registers maintained?
9. Are registers maintained on all vehicles entering and leaving the facility?
10. Are escorts provided when vehicles are permitted access to operating or controlled areas?
11. Are trucks that access operating areas carefully examined?

    Driver
    Contents

VIII. Office Security Checklist

1. Is access controlled?
2. Is the main/visitor entrance monitored by security?
3. Are visitors required to show identification to receptionist?
4. Is there a clear view of entrance, stairs, and elevators from receptionist's desk?
5. Is there a visitor log?
6. Are visitors issued badges?
7. Are visitors escorted in controlled areas?

8. Are the police, fire, and EMS telephone numbers posted?
9. Are bomb threat guidelines posted by each telephone?
10. Are office doors locked after hours or when unattended for a long period of time?
11. Are desks and files locked when office is left unattended?
12. Are items of value left on desks or in an unsecure manner?
13. Is high-value hardware secured? (computers, TV, projectors, etc.)
14. Are filing cabinet keys removed from locks and secured after opening of cabinets for the workday?
15. Is there a lockable drawer in every desk for secure storage of personal effects?
16. Are critical assets secured in a high-security container?
17. What is the fire rating on this container?
18. Are all important papers shredded before discarding in wastebaskets?
19. Are briefcases containing important papers secured when not in use? How?
20. Are desks cleared nightly and documents placed in locked, fireproof safes or cabinets?
21. If people are working after hours, are the exterior entrances locked?
22. Are the financial handling areas separate and secure?
23. Are there clearly established money handling procedures to follow for safeguarding cash, deposits, etc.?

IX. Exterior Physical Characteristics

Perimeter/Grounds

1. What are the physical boundaries of the grounds?
2. Is the perimeter of the grounds identifiable?
3. Is the perimeter clearly marked with signage?
4. Is there a fence?
5. Is the fence strong and in good repair?
6. Fence type:

   Chain-link: Link spacing: ____ Height: ____ Gauge: __
   Other high security
   Decorative

7. Is there a 10-foot clear zone on each side of the fence?
8. Are there any materials within 10 feet of the fence?
9. How close are buildings to the fence?
10. Are unsecured overpasses or subterranean passageways near fence?
11. Are gates solid and in good condition?
12. Are gates locked?
13. Are gates' hinges secure and nonremovable?
14. How are gates secured?
15. How often are locked gates checked?
16. Is the perimeter reinforced by protective lighting? CCTV? IDS?

    Do lights provide light inside and outside the fence?
17. Are there passive features that provide protection? (tiger trap, wall, planters, etc.)

Building Exterior

18. Is the building exterior the first line of defense?
19. What is the construction of the building exterior?
20. Are pedestrian entrances designed to deter vehicle entry?
21. Are crates, stock, or merchandise allowed to be piled near the building?
22. Are blind alleys near buildings protected?

Exterior Doors/Entrances

23. Are entrances:
    Well lit?
    Monitored by CCTV?
    Access controlled?
    Protected by IDS?
24. Are all doors strong and formidable?
    Delay factor: High, medium, low.
25. Are all door frames well constructed and in good condition?
26. Are all door hinge pins located on the inside?
27. Are all door hinges installed so that it would be difficult to remove the door(s)?

28. Are all unused doors secured?
29. Is there at least one lock on each outer door?
30. Are the exterior locks double cylinder, dead bolts, or jimmy-proof-type locks?
31. Can the breaking of glass or a door panel allow the person to open the door?

    Is this glass protected by security glazing or a security screen or bars?

32. Are all locks working properly?
33. Do padlocks, chains, and hasps meet minimum security standards?
34. Are the hasps installed so that the screws/bolts cannot be removed?

Exterior Windows

35. Are windows protected by security glazing or a security screen or bars?
36. Do windows that open have locks that cannot be opened by breaking the glass?
37. Are these windows protected by security glazing or a security screen or bars?
38. Can windows be removed without breaking them?
39. Are windows alarmed?
40. Are small or expensive items left in windows overnight?
41. Are windows located under loading docks or similar structures?

Other Openings

42. Are all vents or utility accesses having a gross area of 1 square foot or more secured with protective coverings?
43. Are fan openings and ventilator shafts protected?
44. Are there manholes, service tunnels, or sewer lines that give direct access to your building/grounds?

    On-site
    Off-site

45. Are openings that are no longer used permanently secured?

46. Are accessible skylights protected with a barrier or an intrusion alarm?
47. Are roof hatches properly secured?
48. Do fire escapes comply with city and state fire regulations?
49. Are your fire exits or escapes designed so that a person can leave easily but would have difficulty in entering?
50. Are dedicated fire exits clearly marked?
51. Are dedicated fire exits clearly alarmed?
52. Are your sidewalk doors/grates locked properly and secured?
53. Are your sidewalk doors/grates securely in place so that the entire frame cannot be pried open?

Exterior Lighting

54. Are the perimeter/grounds lighted?
55. During what hours is lighting used?
56. How much lighting is there for the grounds?

    Type
    Lumens
    Dark spots?
    Support CCTV?
    Lux readings: _

57. How much lighting is there for entrances, guard booths, and sensitive areas?
    Type
    Lumens
    Dark spots?
    Support CCTV?
    Lux readings (at each location): ____

58. Can protective lights be compromised easily (i.e., unscrewing of bulbs)?
59. Are light fixtures vandal-proof?
60. Is the protective lighting system set up on multiple circuits?
61. Is wiring for protective lighting securely mounted?

    Is the wiring in tamper-resistant conduits?
    Is the wiring mounted underground?
    If aboveground, is the wiring high enough to reduce tampering?

62. Are switches and controls properly located, controlled, and protected?
    Are switches weatherproof and tamper resistant?
    Are switches accessible to security personnel?
    Are switches located so that they are inaccessible from outside the perimeter barrier?
    Is there a centrally located switch to control protective lighting? Is it vulnerable?
63. Is there a maintenance schedule for lights? How often are lights checked?
64. Does the facility have a dependable auxiliary source of power?
65. Is there alternate power for the lighting system independent of the facility power system?
66. Is emergency equipment tested periodically?
67. Is emergency equipment designed to go into operation automatically when needed?
68. Is lighting sufficient for CCTV use at night?
69. Are there laws or ordinances that specify a minimum or maximum lux?

X. **Interior Physical Characteristics**

1. List number of rooms occupied by the various departments and officers.
2. List the estimated dollar value of equipment and property in each department/office.
3. Who does maintenance? Contract or proprietary?
4. Does maintenance work during or after business regular hours?
5. What area contains the most sensitive material?
6. Wall material:

   Delay factor: High, medium, low.

7. Do walls extend to the structural ceiling?
8. Interior door material:

   Delay factor: High, medium, low.

9. Interior locks:
   Delay factor: High, medium, low.

Interior Lighting

10. Is adequate lighting for guard use provided on indoor routes?
11. Is there a backup system for emergency lights?
12. Is nighttime lighting sufficient for security purposes?
13. Is lighting sufficient for CCTV?

XI. Intrusion Detection System

1. Are there perimeter sensor systems in use?
   Type: Microwave, infrared, ported cable, fence, seismic, magnetic cable, video motion detection, other: ___.
2. What are concerns about defeat and nuisance alarms?
3. Who monitors the system?
4. Is there an interior IDS?

   Type: ____
   Manufacturer/installer: ____
5. What type of sensors?
   Microwave
   Passive infrared
   Mechanical switch
6. Who monitors?
7. Does the system indicate an alert only within the facility?
8. Does it signal in a central station outside the facility?
9. Is it connected to facility guard headquarters?
10. Is it connected directly to an enforcement headquarters outside the facility proper?
11. Who responds to alarms?

    Guard
    Police
    Other: ____
12. Is there an alarm system protecting high-value storage areas?
13. Is there an alarm system protecting other internal areas?
14. Are there internal or external conditions that may affect the alarm?

15. Is the system inspected and maintained regularly by certified personnel?
16. Are records kept of all alarm signals received, to include time, date, location, action taken, and cause for alarm?
17. Have the causes of alarm malfunctions been remedied?
18. Has the response to an alarm been tested?
19. Have key personnel been given specific alarm control assignments to include alarm opening, closing, checkout procedures, and accountability?
20. Is the alarm system for operating areas turned off during working hours?
21. Is the system tested prior to activating it for nonoperational periods?
22. Is the system tamper resistant?
23. Is there an alternate, or independent, source of power in the event of power failure?
24. Is the emergency power source designed to activate automatically?

XII. Access Control

1. How is the employee entrance to the facility controlled during business hours?
   Open entry
   Keys
   Access cards
   ID badges
   Security officers

2. How is the employee entrance to the facility controlled after hours?
   Open entry
   Keys
   Access cards
   ID badges
   Security officers

Access Control System

3. Is there an access control system?
4. Does this access control system support access levels?
5. Is there control of employee movement between areas within the facility?
6. Are identification badges issued to all employees?
7. Is the wearing of badges required?
8. Is the wearing of badges enforced?
9. Is identification required of all visitors entering the facility?
10. Are all visitors registered?
11. Are employees instructed to challenge strangers in their work areas?
12. Type of system:

    Type of readers:
    Location of readers:
    Multiple system operation? What?

13. Is there a biometric system in use?
14. Are cards, passwords, codes, and other access inputs immediately stopped on an employee termination? Quitting? Transfer?

Locks

15. Are all locks in working order?
16. Are lock bolts protected or constructed so that they cannot be cut?
17. Are locks changed on a regular basis?
18. When was the last time the locks were changed?
19. Are locks changed immediately upon theft or loss of keys?
20. Are locksmith duties responsible by someone on site?
21. Are locksmith duties contracted?
22. Are combinations or keys accessible only to those individuals whose duties require access to them?
23. Are current records kept of combinations to safes and the dates when these combinations are changed?
24. Are these records protected?

25. If combination compromise is suspected, is combination changed immediately?
26. Are locks/combinations changed upon resignation, termination, or suspension of an employee who had a key/combination?
27. Are combination/cypher lock combinations changed annually?
28. Are locks on inactive gates and storage facilities under seal?
29. Are seals checked regularly by supervisory or key control personnel?
30. Are measures in effect to prevent the unauthorized removal of locks on open cabinets, gates, or buildings?

Key Control

31. Who is in charge of key control?
32. Does the key control officer have overall authority and responsibility for issuance and replacement of locks and keys?
33. What is the policy for the issuance of keys?
34. Is the policy in writing?
35. Who determines who is authorized to issue a key?
36. Are keys issued to other than facility personnel?
37. Is the removal of keys from the premises prohibited?
38. Do you keep complete up-to-date records of the disposition of all office keys?
39. Are there procedures that prevent unauthorized personnel from reporting a lost key and receiving a "replacement"?
40. Do you have adequate procedures for collecting keys from former employees?
41. Do you restrict duplication of office keys?
42. Are the keys marked "Do Not Duplicate"?
43. Are inventories and inspections conducted by the key control officer to ensure compliance with directives? How often?
44. When was the last visual key audit made? (to ensure they have not been loaned, lost, or stolen)
45. Are all the keys accounted for? If not, how many are missing?
46. Are all keys systematically stored in a secured wall cabinet of either your own design or from a commercial key control system?
47. Is the key storage cabinet kept locked when not in use?
48. Who has access to the cabinet?

49. Is the key or combination to this cabinet maintained under appropriate security?
50. Are issued keys cross-referenced?
51. Are current records maintained indicating the following:

    Buildings or entrances for which keys are issued?
    Number and identification of keys issued?
    Location and number of master keys?
    Location and number of duplicate keys?
    Issue and turn-in of keys?
    Location of locks and keys held in reverse?
    Clear record of person to whom key is issued?

XIII. CCTV

1. Does this facility have closed-circuit television?
2. Is it monitored continuously? Who monitors?
3. Is it recorded (constant) 24/7?
4. Is it motion/event triggered to record?
5. Is it used for surveillance?
6. Is it used for access control?
7. Recorded to:

    VHS tape
    DVR
    Hard disc

8. Locations of cameras
9. Cameras in use: B&W, color, infrared, PTZ, wireless, RF.
10. Who responds to events? __
11. Is an alternate or independent source of power available for use in the event of power failure?
12. Is the emergency power source designed to cut in and operate automatically?

XIV. Response Force

1. Is there an on-site guard force?
2. Is there an off-site response force? (other than police)

3. What is the response time for off-site forces?

   Police: __
   Other: ___

4. Who provides present guard coverage?
   Proprietary: Manager in charge: ____
   Contract: Contracting official in charge: ___

5. How many total regular hours of duty per week?
6. Is there additional coverage during regular operating hours? During off hours?
7. How many total additional hours per week?
8. How many stationary guard posts exist?
9. How many patrol routes?
10. Are guard assignments and patrol routes varied to avoid a routine?
11. Do guards record their patrol with portable watch clocks or wands?
12. Are general and special orders properly posted?
13. Are guard orders reviewed periodically to ensure applicability?
14. Are duties other than those related to security performed by guard personnel?
15. Are there armed officers at this facility?

    Type of weapons?

16. Are the weapons kept in arms racks and adequately secured when not in use?
17. Are ammunition supplies properly secured and issued only for authorized purposes?
18. Indicate authorized and actual strength, broken down by positions.

    Stationary post
    Patrol
    Armed
    Unarmed
    Other (e.g., K-9)
    Supervisor

19. Have there been changes since the last survey in either the authorized and actual force strength?

20. Is present guard force strength commensurate with the degree of security protection required?
21. Is the use of guard forces reviewed periodically to assure effective and economical use?
22. Do physical, functional, or other changes at the installation indicate the necessity for, or feasibility of:

    Establishing additional guard posts
    Discontinuing any existing posts or patrols

23. Do guards make written reports of incidents?
24. What is the ratio of supervisors to employees?
25. Is supervisory responsibility for guard force operations vested in a security officer?
26. Who does security answer to in management?
27. Is a guard headquarters provided?
28. Does the guard headquarters have direct communications with local municipal fire and police headquarters?
29. What are the standards for the guard force? (knowledge, skill of each task)
30. Are background checks conducted on security personnel?
31. Is each member of the guard force required to complete a course of basic training and take periodic courses of in-service training?
32. Is there a task analysis to identify training needs?
33. Has there been a validation of the task analysis?
34. Does the training cover:

    General and special guard orders
    Preparation of written reports
    Duties in the event of fire, explosion, natural disaster, or civil disturbance
    Location and use of first aid equipment
    Jurisdiction and authority: Search, arrest, force
    Use of lethal and nonlethal force
    Care and use of weapons
    Vehicle operation and safety
    Conditions that may cause fire and explosions
    Types of bombs and explosives
    Location of hazardous materials and processes

Location and use of fire protective equipment, including sprinkler control valves
Location and operation of all important steam and gas valves and main electrical switches
Common forms of pilferage, theft, and sabotage activity
Use of equipment: Communications, vehicles, tour wands, magnetometer, etc.
Observation and description
Patrol
Supervision of visitors
Proper methods of search

XV. **Security Officer Responsibilities and Duties**

Check those that apply:
Visitor control/escort
Employee access control
Personnel and package screening (metal, explosive, and radiation detectors, x-ray and particulate explosive detection systems)
Vehicle gate control/parking lot
Loading dock/inspection of vehicles
Property control
Inspection of vehicles
Key control
Building tours
Operate patrol vehicle(s)
Cash handling
Concierge services
Monitoring environmental conditions
Monitoring alarm system
Monitoring access control systems
Monitoring CCTV systems
Issuing ID badges/access cards
Maintaining/programming systems
Responding to alarm events
Firefighting
Emergency medical response
Bomb scares/emergency evacuation
Other: ___

Index